DevOps 企业级 CI/CD 实战

李泽阳　编著

清华大学出版社
北京

内 容 简 介

本书主要围绕 DevOps 的核心 CI/CD，详细讲解了企业级 CI/CD 的相关技术内容。全书共 11 章，首先讲解了持续集成系统 Jenkins 入门知识、Jenkins 运维管理、Jenkins 流水线的语法与进阶。接着讲解了持续集成阶段的工具平台实践、GitLab 版本控制系统实践、Maven 等构建工具实践、SonarQube 代码质量平台实践、Nexus Repository 3 制品库平台实践。持续部署分别基于云主机和 Kubernetes 环境的持续集成和持续部署实践。最后讲解了使用基础设施即代码工具 Terraform 管理阿里云平台资源。通过本书的学习，读者将熟练掌握企业级 CI/CD 的实践方法和思路。

本书适合具有 1～3 年运维或开发工作经验、或者对 DevOps 感兴趣的读者学习。

本书封面贴有清华大学出版社防伪标签，无标签者不得销售。
版权所有，侵权必究。举报：010-62782989，beiqinquan@tup.tsinghua.edu.cn。

图书在版编目（CIP）数据

DevOps：企业级 CI/CD 实战 / 李泽阳编著. —北京：清华大学出版社，2024.2
ISBN 978-7-302-65185-7

Ⅰ．①D… Ⅱ．①李… Ⅲ．①软件工程 Ⅳ．①TP311.5

中国国家版本馆 CIP 数据核字（2024）第 018802 号

责任编辑：王秋阳
封面设计：秦　丽
版式设计：文森时代
责任校对：马军令
责任印制：丛怀宇

出版发行：清华大学出版社
网　　址：https://www.tup.com.cn，https://www.wqxuetang.com
地　　址：北京清华大学学研大厦 A 座　　邮　编：100084
社 总 机：010-83470000　　邮　购：010-62786544
投稿与读者服务：010-62776969，c-service@tup.tsinghua.edu.cn
质量反馈：010-62772015，zhiliang@tup.tsinghua.edu.cn

印 装 者：三河市东方印刷有限公司
经　　销：全国新华书店
开　　本：186mm×240mm　　印　张：15　　字　数：300 千字
版　　次：2024 年 2 月第 1 版　　印　次：2024 年 2 月第 1 次印刷
定　　价：89.00 元

产品编号：098459-01

前言
Preface

创作背景

DevOps 是一组实践，由人、工具和文化理念组成。DevOps 的核心是实现软件开发团队和 IT 运维团队之间的流程自动化。自 2018 年起，笔者参与了大型企业中多个项目的 DevOps 项目实施和改进，从中积累了丰富的实践经验。于是将实践思考与开发经验整理成一本书分享给同路人共同学习和交流。

DevOps 涵盖的范围非常广，本书主要讲解 DevOps 方法论中的 CI/CD 部分。从理论到实践，分别从持续集成阶段工具和持续部署阶段工具出发和落地。注意：本书中的工具均采用开源版本。

目标读者

- ☑ 运维工程师。
- ☑ 开发工程师。
- ☑ DevOps 工程师。

路线图

本书共 11 章：

- ☑ 第 1 章介绍了持续集成和持续部署的核心工具 Jenkins 的入门知识，包括 Jenkins 持续集成工具的安装部署。
- ☑ 第 2 章分别通过用户管理、权限管理、凭据管理 3 个方面讲解 Jenkins 的系统运维管理。读者可以掌握 Jenkins 运维管理技能。
- ☑ 第 3 章开始讲解 Jenkins 的核心特性流水线即代码实践。读者可以学习 Pipeline 定义、Pipeline 核心语法、Pipeline 开发工具、共享库实践。

- ☑ 第 4 章是对 Pipeline 的进阶实践。读者可以学习基于 Groovy 扩展流水线、流水线触发器、流水线中常用的 DSL 方法。
- ☑ 第 5 章讲解持续集成阶段的代码管理平台实践。读者可以学习 GitLab 代码管理平台实践和项目构建工具实践，了解从源代码管理到编译构建的过程。
- ☑ 第 6 章讲解持续集成阶段的代码质量平台实践。读者可以学习 SonarQube 平台的安装部署、配置管理、与 Jenkins 集成实践。
- ☑ 第 7 章讲解持续集成阶段的制品库平台实践。读者可以学习使用 Nexus Repository 3 集中管理源代码构建制品和依赖，便于一次构建，发布到不同的环境中运行。
- ☑ 第 8 章讲解持续部署阶段的云主机环境下的持续集成和持续部署流水线案例。读者可以学习使用 Jenkins 持续集成生成制品，再通过集成 Ansible 进行批量发布。
- ☑ 第 9 章讲解 Kubernetes 环境持续集成和持续部署相关的基础知识。读者可以学习 Docker 容器、Kubernetes 基础知识，以及常用的发布策略原理，这些是对下一章实践的铺垫。
- ☑ 第 10 章讲解持续部署阶段的 Kubernetes 环境下的持续集成和持续部署流水线案例。读者可以学习应用从源代码构建生成镜像、集成 Helm 发布到 Kubernetes 环境的整个过程。
- ☑ 第 11 章扩展 Terraform 基础设施及代码工具的实践，讲解了 Terraform 工具的实践方式。读者可以学习使用 Terraform 以代码的方式管理阿里云平台资源。

读者服务

- ☑ 示例代码。
- ☑ 学习视频。

读者可以通过扫码访问本书专享资源官网，获取示例代码、学习视频，加入读者群，下载最新学习资源或反馈书中的问题。

勘误和支持

由于笔者水平有限，书中难免会有疏漏和不妥之处，恳请广大读者批评指正。

致谢

　　首先,我要感谢我的家人,他们是我最可靠的后盾,一直在背后支持我,给我鼓励和信心,让我能够专注于本书的撰写。感谢清华大学出版社编辑,她为这本书的撰写和出版做出了巨大的贡献。她的专业知识和敏锐的洞察力帮助我不断改进书稿,使其更加出色。此外,我还要感谢那些曾经为我提供帮助的人,包括我的导师、同事、朋友。他们的支持和鼓励是我前进的动力。最后,我想向所有的读者表示感谢,希望你们能够喜欢这本书,并且从中获得更多的知识和技能。

李泽阳

目 录
Contents

第1章　Jenkins 系统入门 1
  1.1　Jenkins 系统概述 1
     1.1.1　Jenkins 概述 1
     1.1.2　Jenkins 应用场景 2
  1.2　Jenkins 系统安装 3
     1.2.1　准备工作 4
     1.2.2　安装 Jenkins 5
     1.2.3　初始化 7
     1.2.4　安装 Agent 节点 11
  1.3　Jenkins 数据目录 16
  1.4　本章小结 18

第2章　Jenkins 系统管理 19
  2.1　用户管理 19
     2.1.1　Jenkins 本地用户 19
     2.1.2　LDAP 认证集成 21
  2.2　权限管理 24
     2.2.1　准备 24
     2.2.2　安装 Role-based 插件 26
     2.2.3　创建角色 27
     2.2.4　授权角色 29
     2.2.5　测试权限 29
  2.3　凭据管理 30
     2.3.1　安装凭据插件 30
     2.3.2　创建凭据 32
  2.4　本章小结 33

第3章　Jenkins Pipeline 实战 34
  3.1　什么是 Pipeline 34
     3.1.1　Pipeline 简介 34
     3.1.2　为什么使用 Pipeline 35
     3.1.3　什么是 Jenkinsfile 36
  3.2　Pipeline 核心语法 36
     3.2.1　agent 节点 37
     3.2.2　stages 阶段 38
     3.2.3　post 构建后操作 38
     3.2.4　environment 环境变量 39
     3.2.5　options 运行选项 40
     3.2.6　parameters 参数 41
     3.2.7　triggers 触发器 42
     3.2.8　input 交互 43
     3.2.9　when 阶段运行控制 44
     3.2.10　parallel 并行运行 45
  3.3　Pipeline 开发工具 46
  3.4　共享库实践 49
     3.4.1　创建共享库 49
     3.4.2　编写共享库代码 50
     3.4.3　修改全局设置 50
     3.4.4　加载共享库 52
  3.5　本章小结 53

第4章　Jenkins Pipeline 进阶 54
  4.1　Groovy 编程语法 54

4.1.1	数据类型	54
4.1.2	控制语句	58
4.1.3	异常处理	61
4.1.4	函数	61

4.2 Jenkins 触发器 62
- 4.2.1 安装触发器 63
- 4.2.2 配置触发器 63
- 4.2.3 解析 Request 参数 65
- 4.2.4 解析 Header 参数 68
- 4.2.5 解析 Post 参数 70

4.3 常用的 DSL 语句 73
- 4.3.1 获取当前触发用户 73
- 4.3.2 JSON 数据解析 75
- 4.3.3 在 Pipeline 中使用凭据 76
- 4.3.4 自定义构建 ID 和描述 77

4.4 本章小结 78

第 5 章 项目代码管理 79

5.1 GitLab 系统入门 79
- 5.1.1 GitLab 概述 79
- 5.1.2 GitLab 安装部署 80

5.2 GitLab 工作流 82
- 5.2.1 创建项目组和项目 82
- 5.2.2 生成和提交项目代码 86
- 5.2.3 分支开发策略 87

5.3 提交流水线实践 89
- 5.3.1 Jenkins 配置 89
- 5.3.2 GitLab 配置 92
- 5.3.3 编写 Pipeline 96
- 5.3.4 Pipeline 优化 100

5.4 项目构建工具 102
- 5.4.1 Maven 构建 102
- 5.4.2 Gradle 构建 104

- 5.4.3 NPM 构建 106

5.5 本章小结 108

第 6 章 代码质量平台实战 109

6.1 SonarQube 系统入门 109
- 6.1.1 SonarQube 概述 109
- 6.1.2 SonarQube 安装 111
- 6.1.3 插件管理 113

6.2 SonarQube 代码扫描 115
- 6.2.1 SonarQube 质量配置 115
- 6.2.2 SonarQube 质量阈 116
- 6.2.3 Sonar Scanner 配置 117

6.3 SonarQube 系统集成 121
- 6.3.1 准备工作 121
- 6.3.2 命令行方式 122
- 6.3.3 Jenkins 插件 127
- 6.3.4 多分支代码扫描 130

6.4 本章小结 134

第 7 章 制品库平台实战 135

7.1 制品库平台入门 135
- 7.1.1 管理规范 135
- 7.1.2 Nexus Repository 3 概述 136

7.2 Nexus Repository 实践 137
- 7.2.1 Nexus Repository 3 安装 137
- 7.2.2 搭建 Maven 私服仓库 138
- 7.2.3 搭建 Maven 本地仓库 140
- 7.2.4 制品上传方式 141

7.3 Nexus Repository 扩展实践 144
- 7.3.1 调试 REST API 144
- 7.3.2 上传 Raw 类型制品 146
- 7.3.3 Jenkins 插件上传制品 147

7.4 本章小结 149

目录

第 8 章　云主机环境持续部署............150
8.1　项目准备工作............................150
8.1.1　分支策略............................150
8.1.2　环境准备............................151
8.1.3　Ansible 配置.......................152
8.1.4　Pipeline 设计......................153
8.2　持续集成实践............................154
8.2.1　准备工作............................154
8.2.2　设置 Pipeline......................157
8.3　持续部署实践............................164
8.3.1　准备工作............................164
8.3.2　设置 Pipeline......................167
8.4　本章小结...................................173

第 9 章　Kubernetes 基础..................175
9.1　Docker 容器基础.......................175
9.1.1　Docker 简介........................175
9.1.2　Docker 镜像构建.................176
9.1.3　Docker 镜像管理.................178
9.2　Kubernetes 基础........................180
9.2.1　资源对象............................181
9.2.2　Kubectl 工具发布.................183
9.2.3　Helm 工具发布....................183
9.3　Kubernetes 部署策略..................184
9.3.1　滚动更新............................184
9.3.2　零停机部署........................185
9.4　本章小结...................................186

第 10 章　Kubernetes 持续部署..........187
10.1　持续集成流水线........................187
10.1.1　准备工作...........................187
10.1.2　设置 Pipeline.....................190
10.1.3　启用 GitOps.......................196
10.2　基于 Kubectl 持续部署.............205
10.2.1　准备工作...........................205
10.2.2　设置 Pipeline.....................207
10.3　基于 Helm 持续部署................211
10.3.1　准备工作...........................212
10.3.2　设置 Pipeline.....................213
10.4　本章小结..................................218

第 11 章　基础设施即代码.................219
11.1　Terraform 入门........................219
11.2　供应商 Provider.......................221
11.3　定义云资源..............................224
11.4　开通资源.................................226
11.5　本章小结.................................229

第 1 章
Jenkins 系统入门

众所周知，CI（持续集成）工具对于 DevOps 来说是不可或缺的。目前，CI 工具有多种选择，Jenkins 是最受欢迎的工具之一。随着 DevOps 在国内的发展，企业在建设 DevOps 工具链的过程中开始广泛采用 Jenkins 作为持续集成和持续交付引擎。Jenkins 可以将我们的日常工作流程自动化和可视化编排，以提升工作效率，减少人为操作带来的错误。本章开始，笔者将汇总自己的实践经验，为读者展示如何快速入门 Jenkins 系统。

本章内容速览。
- ☑ Jenkins 系统概述
- ☑ Jenkins 系统安装
- ☑ Jenkins 数据目录
- ☑ 本章小结

1.1 Jenkins 系统概述

Jenkins 是企业广泛应用的开源持续集成和持续部署解决方案，它具有丰富的插件资源，可以在不同公司环境下进行功能扩展。本节将介绍 Jenkins 相关的概念以便于初学者快速入门。

1.1.1 Jenkins 概述

Jenkins 是基于 Java 语言开发的开源持续集成系统。它提供了数千个插件来支持构建、部署和自动化任何项目，并提供了一种简单的方法来设置持续集成和持续部署流水线。其插件 Pipeline（流水线）可以为任何语言和源代码仓库的项目建立流水线，也可以自动化许多其他日常开发任务。Jenkins 在 DevOps 实践中提供了一种更快、更强大的方式来集成整个构建、测试和部署工具链。Jenkins 工具链集成图如图 1-1 所示。

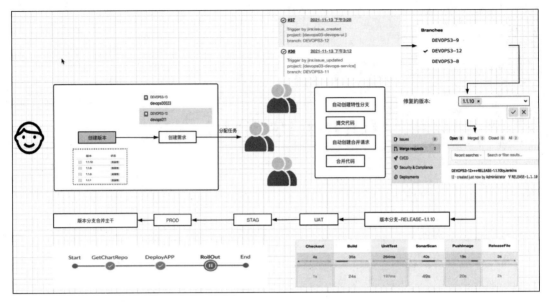

图 1-1 Jenkins 工具链集成图

1.1.2 Jenkins 应用场景

Jenkins 可用于自动执行与构建、测试和交付或部署软件相关的各种任务。在软件工程领域，使用 Jenkins 核心特性 Pipeline 构建持续集成和持续部署流水线（见图 1-2）。Jenkins 的应用场景非常广，下面笔者从开发、运维、测试的角度分别举例说明。

图 1-2 持续集成与持续部署流水线图

1. 持续集成

开发人员的主要工作是编写、验证、提交业务代码。在没有使用 Jenkins 之前，这些验证任务都是手动运行的，随着提交代码频率的增加，可能会出现验证不完整，导致很多集成错误。使用 Jenkins 自动化并开启构建触发器能够集成版本控制系统，能实现代码提交到版本控制系统之后，自动运行代码验证并尽快发现集成错误，尽早反馈。持续集成工作流如图 1-3 所示。

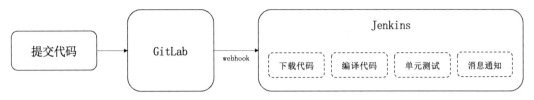

图 1-3　持续集成工作流

2. 持续部署

运维人员的工作主要是将开发的应用程序部署到各个环境中。在没有使用 Jenkins 之前，这些操作都是通过手动或者编写 Shell 脚本实现的，很容易出现人工错误、发布持续时间也不太稳定的情况。使用 Jenkins 之后，可以将脚本集成到 Jenkins Pipeline，运行时将脚本的日志和步骤进行可视化展示。通过参数化构建复用发布步骤，传递环境参数进行不同环境的自动化发布。持续部署工作流如图 1-4 所示。

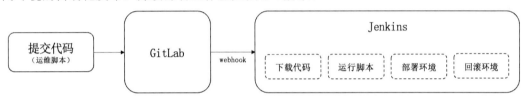

图 1-4　持续部署工作流

3. 持续测试

测试人员的工作是编写一些接口和 Web UI 等方面的测试用例。在没有使用 Jenkins 之前，测试是在本地通过命令行运行的。使用 Jenkins 之后可以将测试用例和相关脚本代码存储到版本控制系统，通过 Jenkins 自动化运行测试用例并使用测试相关的插件将测试结果展示到页面。持续测试工作流如图 1-5 所示。

图 1-5　持续测试工作流

1.2　Jenkins 系统安装

Jenkins 是基于 Java 语言开发的持续集成系统，可以运行在任何安装了 Java 运行环境的计算机上。本节，笔者将为读者逐步演示 Jenkins 服务端（Server）和代理端（Agent）的安装部署。

1.2.1 准备工作

Jenkins 采用分布式架构，由服务端节点和代理端节点组成。服务端主要负责流水线作业的调度，代理端实际运行流水线作业。Jenkins 架构图如图 1-6 所示。

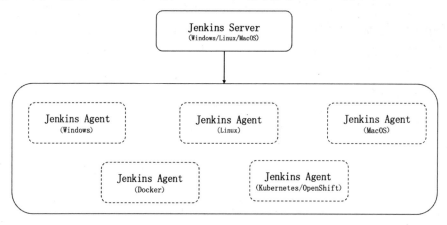

图 1-6　Jenkins 架构图

提示：

为了保障 Jenkins 平台的稳定性，笔者建议不要在生产环境的 Jenkins 服务端节点上运行流水线作业——当运行大量任务时，可能会使 Jenkins 节点崩溃，这会影响节点的性能。

1. 操作系统

Jenkins 是跨平台（Windows/Linux/macOS）的，部署方式可以选择通过本机系统包、Docker、Kubernetes 等。笔者实验环境是 Linux CentOS 8 操作系统。以下是系统和内核版本的信息。

```
[root@jenkins-service ~]#cat /etc/redhat-release
CentOS Stream release 8
[root@jenkins-service ~]#uname -r
4.18.0-373.el8.x86_64
```

2. 安装 JDK

Jenkins 是使用 Java 语言开发的，运行时需要依赖 JDK 环境。自 Jenkins LTS（长期支持版本）2.346.x 之后，安装部署时最低版本要求是 JDK11（参考更新说明见 https://www.jenkins.io/doc/upgrade-guide/2.346/）。

JDK11 的安装过程可以参考以下代码。

```
#下载 JDK11 压缩包
[root@jenkins-service ~]#wget https://github.com/adoptium/temurin10-binaries/releases/download/jdk-11.0.16.1%2B1/OpenJDK11U-jdk_x64_linux_hotspot_11.0.16.1_1.tar.gz

#解压
[root@jenkins-service ~]#tar zxf OpenJDK11U-jdk_x64_linux_hotspot_11.0.16.1_1.tar.gz -C /usr/local
[root@jenkins-service ~]# cd /usr/local/jdk-11.0.16.1+1/
[root@jenkins-service jdk-11.0.16.1+1]# ls
bin  conf  include  jmods  legal  lib  man  NOTICE  release

#设置环境变量
[root@jenkins-service ~]#vim /etc/profile
export JAVA_HOME=/usr/local/jdk-11.0.16.1+1
export PATH=$JAVA_HOME/bin:$PATH
[root@jenkins-service ~]#source /etc/profile

#配置软链接
[root@jenkins-service jdk-11.0.16.1+1]# ln -sf /usr/local/jdk-11.0.16.1+1/bin/java /usr/bin/

#验证版本信息
[root@jenkins-service jdk-11.0.16.1+1]# /usr/bin/java -version
openjdk version "11.0.16.1" 2021-07-12
OpenJDK Runtime Environment Temurin-11.0.16.1+1 (build 11.0.16.1+1)
OpenJDK 63-Bit Server VM Temurin-11.0.16.1+1 (build 11.0.16.1+1, mixed mode)
```

1.2.2 安装 Jenkins

在 CentOS 中需要下载系统 rpm 包来安装 Jenkins 服务端。笔者本次安装采用的是 2.346.2-1.1 版本，建议使用国内源下载对应的 rpm 包，如清华源（https://mirrors.tuna.tsinghua.edu.cn/jenkins/）。具体的安装步骤及参考代码如下。

1. 下载系统包

下载系统包的命令如下。

```
[root@jenkins-service~]#wget https://mirrors.tuna.tsinghua.edu.cn/
```

```
jenkins/redhat-stable/jenkins-2.346.3-1.1.noarch.rpm
```

2. 安装系统包

安装系统包的命令如下（见图1-7）。

```
[root@jenkins-service ~]#rpm -ivh jenkins-2.346.3-1.1.noarch.rpm
```

```
[root@jenkins-service ~]# rpm -ivh jenkins-2.346.3-1.1.noarch.rpm
警告：jenkins-2.346.3-1.1.noarch.rpm: 头V4 RSA/SHA512 Signature, 密钥 ID 45f2c3d5: NOKEY
Verifying...                          ################################# [100%]
准备中...                             ################################# [100%]
        软件包 jenkins-2.346.3-1.1.noarch 已经安装
[root@jenkins-service ~]#
```

图1-7　安装 Jenkins 服务

3. 启动服务

下面启动 Jenkins 服务并检查服务的状态，如图1-8所示。

```
[root@jenkins-service ~]# systemctl start jenkins
[root@jenkins-service ~]# systemctl status jenkins
● jenkins.service - Jenkins Continuous Integration Server
   Loaded: loaded (/usr/lib/systemd/system/jenkins.service; enabled; vendor preset: disabled)
   Active: active (running) since Wed 2022-12-14 09:38:21 HKT; 45min ago
 Main PID: 3286 (java)
    Tasks: 53 (limit: 203440)
   Memory: 3.1G
   CGroup: /system.slice/jenkins.service
           └─3286 /usr/bin/java -Djava.awt.headless=true -jar /usr/share/java/jenkins.war --webroot=

12月 14 09:38:21 jenkins-service jenkins[3286]: 2022-12-14 01:38:21.096+0000 [id=24]     INFO
12月 14 09:38:21 jenkins-service systemd[1]: Started Jenkins Continuous Integration Server.
12月 14 09:38:34 zeyang-nuc-service jenkins[3286]: 2022-12-14 01:38:34.328+0000 [id=64]   INFO
12月 14 09:38:34 zeyang-nuc-service jenkins[3286]: 2022-12-14 01:38:34.329+0000 [id=64]   INFO
12月 14 09:38:34 zeyang-nuc-service jenkins[3286]: 2022-12-14 01:38:34.331+0000 [id=64]   INFO
12月 14 09:54:08 zeyang-nuc-service jenkins[3286]: 2022-12-14 01:54:08.862+0000 [id=89]   INFO
12月 14 09:54:08 zeyang-nuc-service jenkins[3286]: 2022-12-14 01:54:08.901+0000 [id=90]   INFO
12月 14 10:04:56 zeyang-nuc-service jenkins[3286]: 2022-12-14 02:04:56.003+0000 [id=111]  INFO
12月 14 10:11:41 zeyang-nuc-service jenkins[3286]: 2022-12-14 02:11:41.784+0000 [id=112]  INFO
12月 14 10:11:41 zeyang-nuc-service jenkins[3286]: 2022-12-14 02:11:41.785+0000 [id=112]  INFO
[root@jenkins-service ~]#
```

图1-8　启动 Jenkins 服务

4. 设置服务开机自启

设置 Jenkins 服务开机自启，如图1-9所示。

```
[root@jenkins-service ~]# systemctl enable jenkins
Synchronizing state of jenkins.service with SysV service script with /usr/lib/systemd/systemd-sysv-in
stall.
Executing: /usr/lib/systemd/systemd-sysv-install enable jenkins
Created symlink /etc/systemd/system/multi-user.target.wants/jenkins.service → /usr/lib/systemd/system
/jenkins.service.
```

图1-9　设置 Jenkins 服务开机自启

当我们完成上面的操作后,可以通过 ps aux 命令查看服务进程。如下面代码所示,通过进程的启动参数可以得知 Jenkins 默认的 HTTP 监听端口是 8080。

```
#查看服务进程
[root@jenkins-service ~]#ps aux | grep java | grep jenkins
jenkins    3779 29.5  9.1 14278716 2984916 ?     Ssl  21:12   0:25 /usr/bin/
java -Djava.awt.headless=true -jar /usr/share/java/jenkins.war --webroot=
/var/cache/jenkins/war --httpPort=8080
```

1.2.3 初始化

Jenkins 服务端安装好之后还有一个初始化的过程,通过浏览器访问 http://<service_ip>:8080 即可进入初始化页面。

初始化的步骤如下。

1. 解锁系统

为了安全地安装 Jenkins,会将解锁需要的密码存储到 Jenkins 启动日志或者/var/lib/jenkins/secrets/initialAdminPassword 文件中。读者可任选其中一种方式,将密码粘贴到输入框,单击"继续"按钮,如图 1-10 所示。

图 1-10　Jenkins 解锁

2. 安装插件

Jenkins 的功能扩展都来自于插件。Jenkins 插件资源丰富,为了集成一些工具,我们不得不安装很多插件,但是在生产环境中,笔者不建议大家安装太多无用的插件,因为插件增多会影响 Jenkins 的启动速度,插件安装方式如图 1-11 所示。

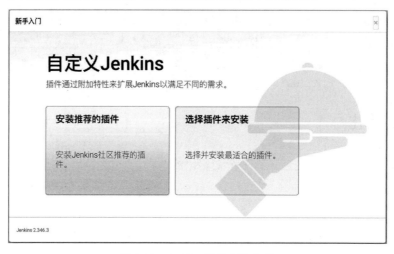

图 1-11　Jenkins 插件安装方式

此处选择"选择插件来安装"进入插件的选择页面,如图 1-12 所示。可以看到 Jenkins 默认选择了一些插件,这里我们单击"无"取消默认选择的插件,然后根据需求安装本地化语言插件。

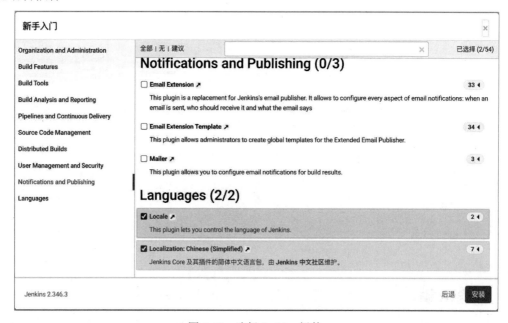

图 1-12　选择 Jenkins 插件

单击"安装"按钮即可进入插件安装进度页面,在该页面可以看到已选择插件的安装进度和日志信息。

> **提示:**
> 这个步骤很容易出现插件安装停顿的问题,多数情况是因为网络问题导致的,因此要保持网络畅通。

3. 创建用户

插件安装完成后,会自动跳转到管理员用户创建页面,填写用户名、密码等信息,单击"保存并完成"按钮,如图 1-13 所示。

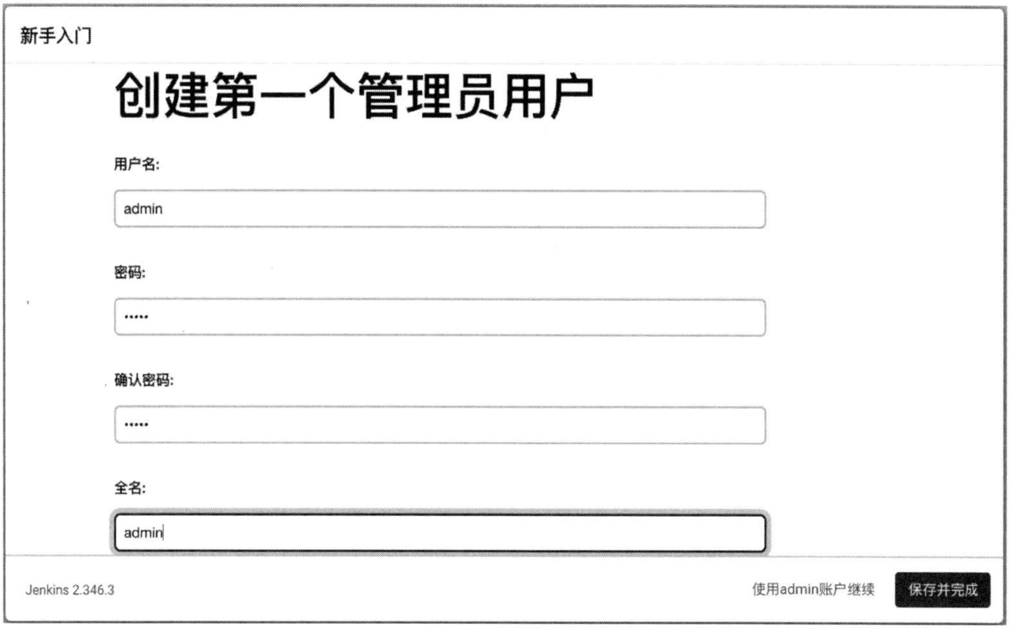

图 1-13 创建管理员用户

4. 配置实例

当前笔者的环境机器 IP 是 192.168.1.200,端口号使用的是 Jenkins 默认设置的 8080。如果读者的端口或者 IP 地址发生了变化,记得更新这里的设置,如图 1-14 所示。

5. 完成初始化

如果进入如图 1-15 所示的页面,意味着安装和初始化 Jenkins 服务端完成。

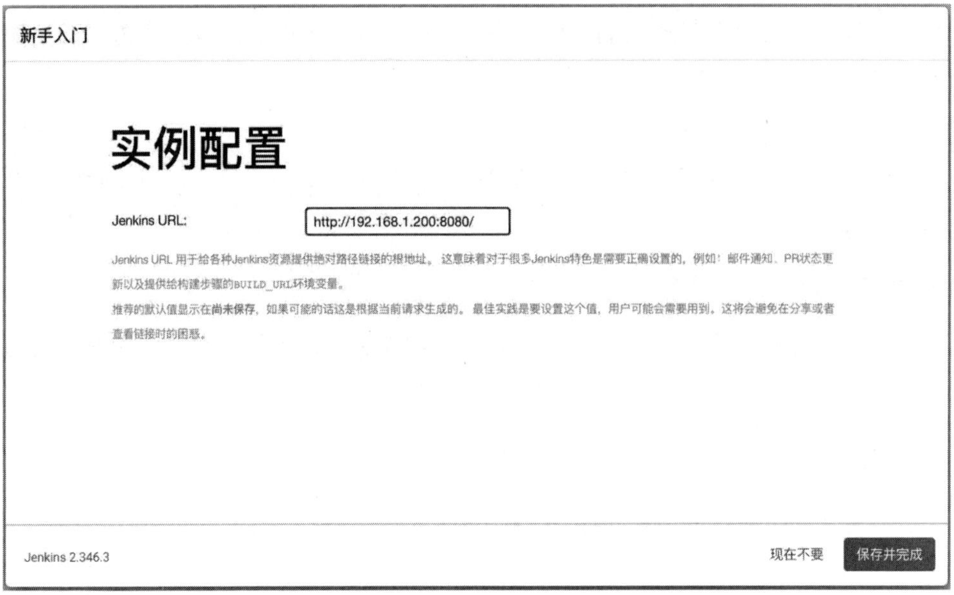

图 1-14 Jenkins 实例配置

图 1-15 Jenkins 安装完成

第一次安装部署完成后，如果 Jenkins 管理页面显示的是英文，重启后会显示中文，如图 1-16 所示。

第 1 章　Jenkins 系统入门

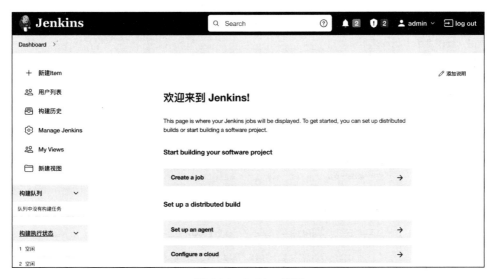

图 1-16　Jenkins 首页

1.2.4　安装 Agent 节点

Jenkins 的代理端主要用于运行 Jenkins 服务端所调度的流水线作业。它的部署方式与服务端一样是跨平台的。为了减少实验环境的资源，笔者将 Jenkins 服务端和代理端部署在同一台机器中。注意：实际生产环境要采用高性能的服务器进行独立部署。

1. 配置节点

单击左侧"系统管理"进入"管理 Jenkins"页面，在页面中可以看到"节点管理"选项，如图 1-17 所示。

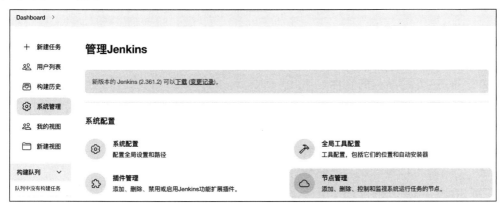

图 1-17　Jenkins 节点管理

进入节点管理页面，可以看到已经在线的服务端 master 节点。单击左侧菜单中的"新建节点"，如图 1-18 所示。

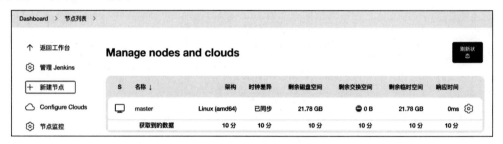

图 1-18　Jenkins 新建节点

进入新建节点页面后，填写要创建的节点名称和 Type（节点类型）。因为笔者将 Jenkins Agent 节点部署到了一台固定的机器上，所以 Type 选择固定节点，如图 1-19 所示。

图 1-19　填写 Jenkins 节点信息

单击 Create 按钮创建节点并进入节点的详细配置页，如图 1-20 所示。

各个配置项的说明如下。

- ☑ 名字：填写节点的名称。
- ☑ 描述：填写节点的描述信息。
- ☑ Number of executors：执行器的数量，默认每一个 Jenkins 作业会分配一个执行器。
- ☑ 远程工作目录：Agent 节点的工作目录。
- ☑ 标签：可以对节点进行分组，便于流水线调度。
- ☑ 用法：哪些作业可以使用这些节点，选择"只允许运行绑定到这台机器的 Job"或者"经常使用该节点"即可运行任意作业。

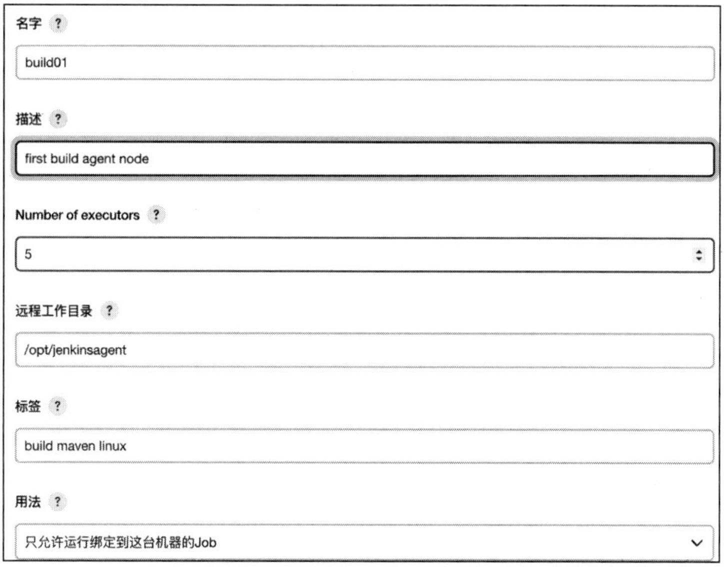

图 1-20　Jenkins 节点详细配置

在 Jenkins 的生产环境中，常用的启动方式为 SSH 和 JNLP。JNLP 对应中文为"通过 Java Web 启动代理"。这里笔者选择"通过 Java Web 启动代理"，如图 1-21 所示。

图 1-21　Jenkins 节点启动方式

DevOps 企业级 CI/CD 实战

> 📖 **提示：**
>
> 如果遇到 "Either WebSocket mode is selected, or the TCP port for inbound agents must be enabled" 错误提示，是因为在 Jenkins 新版本中，默认禁用了 TCP 端口。我们需要导航到 "系统管理 > 管理 Jenkins > 安全 > 全局安全" 配置开启 TCP 端口，如图 1-22 所示。

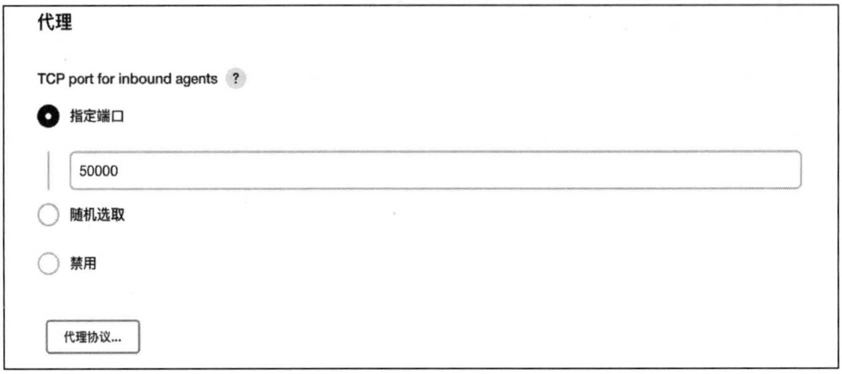

图 1-22　Jenkins Server 与 Agent 通信端口配置

这里笔者默认配置的是 50000 端口，此端口作为后续 Agent 节点启动时与 Server 节点通信使用。

填写好上面的配置并保存。系统会自动进入节点页面，可以看到刚刚增加的节点配置成功（此时节点添加成功但是还不能运行作业，需要启动代理程序），如图 1-23 所示。

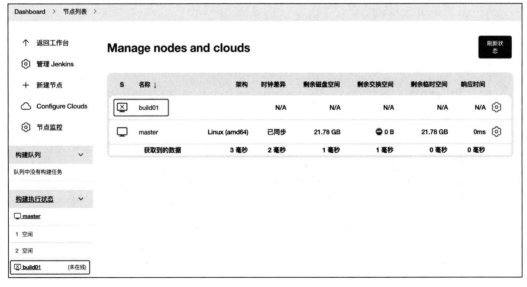

图 1-23　Jenkins 节点管理

2. 启动节点

以“Java Web”方式设置的代理节点，需要下载启动程序。参考图 1-23 单击节点进入节点信息页面，获取 agent.jar 的下载链接。启动节点的步骤如下。

1）创建 Agent 目录

创建一个目录用于存储 Jenkins Agent 的安装程序，命令如下。

```
#创建 Agent 目录
[root@ jenkins-service ~]#mkdir jenkinsagent
[root@ jenkins-service ~]#cd jenkinsagent/
```

2）下载 Agent 启动程序

Agent 启动程序是一个 Java 语言开发的 jar 包。下载的命令如下。

```
#下载代理程序
[root@ jenkins-service jenkinsagent]#wget http://192.168.1.200:8080/
jnlpJars/agent.jar
--2021-9-10 22:27:30--  http://192.168.1.200:8080/jnlpJars/agent.jar
```

3）启动 Agent 程序

使用 Java 命令启动 Agent 程序，如图 1-24 所示。

```
root@jenkins-service:~
[root@jenkins-service jenkinsagent]# java -jar agent.jar -jnlpUrl http://192.168.1.200:8080/computer/
build01/jenkins-agent.jnlp -secret 5497a101324ce5961a07afbb4dffaae3010d598fe7eae7893e37fddaa2fbd366 -
workDir "/opt/jenkinsagent"
12月 14, 2022 9:54:08 上午 org.jenkinsci.remoting.engine.WorkDirManager initializeWorkDir
信息: Using /opt/jenkinsagent/remoting as a remoting work directory
12月 14, 2022 9:54:08 上午 org.jenkinsci.remoting.engine.WorkDirManager setupLogging
信息: Both error and output logs will be printed to /opt/jenkinsagent/remoting
12月 14, 2022 9:54:08 上午 hudson.remoting.jnlp.Main createEngine
信息: Setting up agent: build01
12月 14, 2022 9:54:08 上午 hudson.remoting.jnlp.Main$CuiListener <init>
信息: Jenkins agent is running in headless mode.
12月 14, 2022 9:54:08 上午 hudson.remoting.Engine startEngine
信息: Using Remoting version: 4.13.3
12月 14, 2022 9:54:08 上午 org.jenkinsci.remoting.engine.WorkDirManager initializeWorkDir
信息: Using /opt/jenkinsagent/remoting as a remoting work directory
12月 14, 2022 9:54:08 上午 hudson.remoting.jnlp.Main$CuiListener status
信息: Locating server among [http://192.168.1.200:8080/]
12月 14, 2022 9:54:08 上午 org.jenkinsci.remoting.engine.JnlpAgentEndpointResolver resolve
信息: Remoting server accepts the following protocols: [JNLP4-connect, Ping]
12月 14, 2022 9:54:08 上午 hudson.remoting.jnlp.Main$CuiListener status
信息: Agent discovery successful
  Agent address: 192.168.1.200
  Agent port:    50000
  Identity:      67:fc:14:a8:98:94:c8:67:74:89:3c:2f:5a:2b:8f:84
12月 14, 2022 9:54:08 上午 hudson.remoting.jnlp.Main$CuiListener status
信息: Handshaking
12月 14, 2022 9:54:08 上午 hudson.remoting.jnlp.Main$CuiListener status
信息: Connecting to 192.168.1.200:50000
12月 14, 2022 9:54:08 上午 hudson.remoting.jnlp.Main$CuiListener status
信息: Trying protocol: JNLP4-connect
12月 14, 2022 9:54:08 上午 org.jenkinsci.remoting.protocol.impl.BIONetworkLayer$Reader run
信息: Waiting for ProtocolStack to start.
12月 14, 2022 9:54:09 上午 hudson.remoting.jnlp.Main$CuiListener status
信息: Remote identity confirmed: 67:fc:14:a8:98:94:c8:67:74:89:3c:2f:5a:2b:8f:84
12月 14, 2022 9:54:09 上午 hudson.remoting.jnlp.Main$CuiListener status
信息: Connected
```

图 1-24　启动 Jenkins Agent 程序

> **提示：**
>
> 根据笔者的经验，此步骤出现问题的概率很高。大部分情况都是网络问题导致的，例如检查 Agent 节点与 Jenkins Server 节点的 50000 端口和 8080 端口的连通性是否正常。

当出现 "Connected" 关键字时查看节点管理页面，可以看到该 Agent 已经连接成功。刷新节点页面，如图 1-25 所示。

图 1-25　Jenkins 节点状态

> **提示：**
>
> 当前节点启动成功，但是 Jenkins Agent 程序并未后台运行。也就是说，终端关闭后程序就停止了，需要进一步优化。笔者习惯将启动命令写到脚本中，用 nohup 的方式启动程序可以参考以下代码。

```bash
#!/bin/bash

nohup java -jar agent.jar -jnlpUrl http://192.168.1.200:8080/computer/
build01/jenkins-agent.jnlp -secret @secret-file -workDir "/opt/
jenkinsagent" &
```

至此，我们完成了 Jenkins Agent 节点的安装，我们的基础环境准备好了。

1.3　Jenkins 数据目录

当 Jenkins 运行时，会将其所有的数据（系统日志、插件配置、作业配置等）存储到数据目录中。

Jenkins 数据以 XML 文件格式存储在本地文件系统中，JENKINS_HOME 变量定义了具体的目录位置。如需自定义数据目录，可以重新定义 JENKINS_HOME 变量。数据目录中的内容如下所示。

```
#查看$JENKINS_HOME 目录
[root@jenkins-service jenkins]# ls -l $JENKINS_HOME | awk '{print $NF}'
56
config.xml                                          #Jenkins 系统配置文件
hudson.model.UpdateCenter.xml                       #Jenkins 插件更新源配置文件
identity.key.enc                                    #标识 Jenkins 实例唯一
jenkins.install.InstallUtil.lastExecVersion         #Jenkins 当前版本（插件安装时
会读取这里的版本号）
jenkins.install.UpgradeWizard.state                 #Jenkins 当前的状态
jenkins.model.JenkinsLocationConfiguration.xml      #Jenkins 本地化配置
jenkins.security.apitoken.ApiTokenPropertyConfiguration.xml    #Jenkins
ApiToken 配置
jenkins.security.QueueItemAuthenticatorConfiguration.xml    # Queue 认证配置
jenkins.security.UpdateSiteWarningsConfiguration.xml    #更新站点警告配置
jenkins.telemetry.Correlator.xml                    #Jenkins 数据收集配置
jobs
logs
nodeMonitors.xml
nodes
plugins
queue.xml.bak
secret.key
secret.key.not-so-secret
secrets
updates
userContent
users
```

Jenkins 部分目录的用途如下。

- ☑ jobs：Jenkins 项目。
- ☑ nodes：Jenkins 节点信息。
- ☑ secrets：密钥信息。
- ☑ userContent：类似于 Web 站点目录，可以上传一些文件。
- ☑ logs：日志信息。
- ☑ plugins：插件相关配置。
- ☑ updates：插件更新目录。
- ☑ users：Jenkins 系统用户目录。

> **提示：**
> 在修改 JENKINS_HOME 变量进行数据目录切换时，记得把之前的数据复制到新的数据目录，否则下次 Jenkins 启动时会重新初始化安装并生成新的数据目录。

1.4 本章小结

本章主要讲解了 DevOps 核心中的持续集成系统 Jenkins 的快速入门知识。为读者介绍了 Jenkins 在企业级实践中的应用场景、Jenkins 系统的安装部署、Jenkins 系统各个数据目录的作用。至此，读者已经打开了 DevOps 工程技术的第一扇大门。

下一章将讲解持续集成系统 Jenkins 的系统管理，结合笔者的工作经验为读者介绍企业中一些常用的 Jenkins 的管理实践。

第 2 章
Jenkins 系统管理

基于 Jenkins 建设企业级的持续集成平台还需要掌握一些必备的系统设置。例如，面对企业大规模的用户如何实现账号统一认证、多租户的场景下如何授予用户权限、如何与其他平台集成密钥信息的管理等。本章，笔者将逐一展示企业常用的 Jenkins 系统管理实践。

本章内容速览。
- ☑ 用户管理
- ☑ 权限管理
- ☑ 凭据管理
- ☑ 本章小结

2.1 用 户 管 理

Jenkins 的系统用户来源支持多种方式。当企业团队人数较少时，可以使用本地系统用户的方式进行维护。但在大规模用户场景下，本地系统用户的方式会存在许多问题，例如，维护不便、账号密码跨平台无法同步等。接下来，笔者将根据实践经验为读者展示 Jenkins 本地用户的管理方式，并通过 Jenkins 和 LDAP 认证系统集成来实现用户统一登录。

2.1.1　Jenkins 本地用户

在初始化 Jenkins 的过程中创建了第一个管理员用户。导航到 Jenkins 全局安全配置，可以看到 Jenkins 使用默认的"Jenkins 专有用户数据库"来进行用户管理。Jenkins 全局安全配置如图 2-1 所示。

导航到管理用户页面可以看到当前系统中的所有用户，如图 2-2 所示。

本地用户列表如图 2-3 所示。

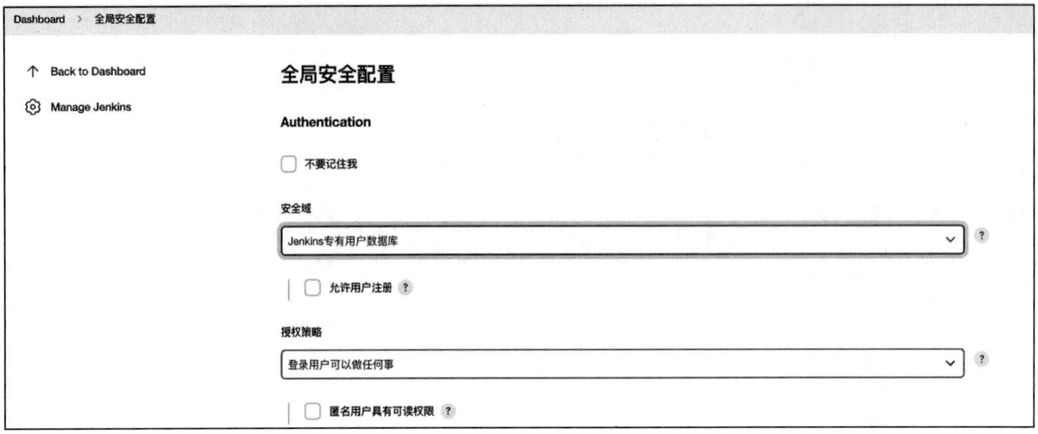

图 2-1　Jenkins 全局安全配置

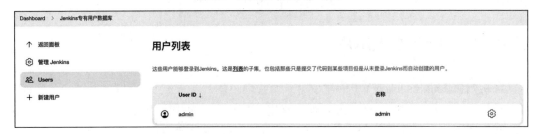

图 2-2　Jenkins 管理用户菜单

图 2-3　本地用户列表

单击左侧"新建用户",填写用户名、密码、确认密码、全名,然后单击"新建用户"按钮完成用户新建,如图 2-4 所示。

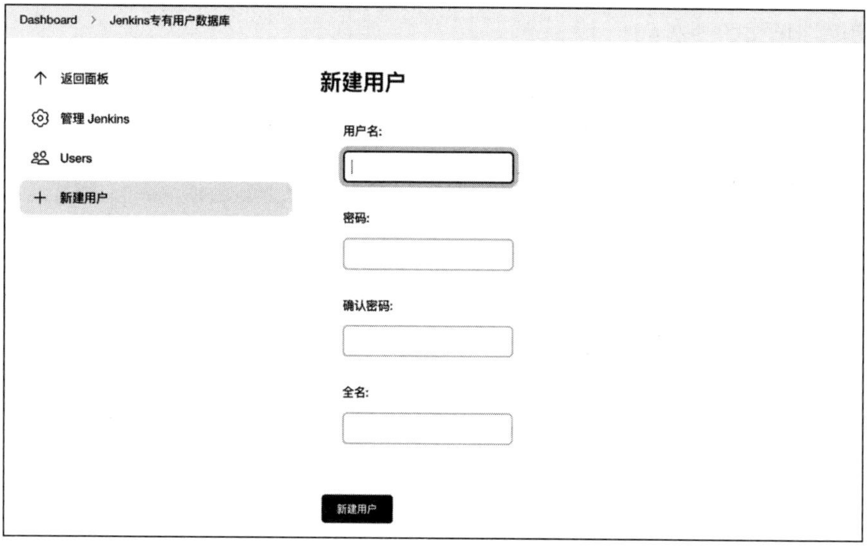

图 2-4　新建用户页面

2.1.2　LDAP 认证集成

LDAP（lightweight directory access protocol，轻量级目录访问协议）是一种成熟、灵活且支持良好的、基于标准的与目录服务器交互的机制。它通常用于身份验证和存储有关用户、组和应用程序的信息。

企业普遍采用 LDAP 系统管理组织中的所有用户。Jenkins 可以通过插件与 LDAP 系统集成，实现用户登入。接下来笔者将演示集成的过程。

1. 获取 LDAP 组织信息

我们需要获取 LDAP 集成所需要的参数，如 LDAP 服务器地址、用户组织信息、具有查询权限的用户。LDAP 系统组织页面如图 2-5 所示。

笔者的环境参数如下。

- ☑ LDAP 服务器：ldap://192.168.1.200:389。
- ☑ 管理员用户：cn=admin,dc=example,dc=com。
- ☑ 用户组织：ou=jenkins,dc=example,dc=com。
- ☑ 测试用户：cn=test1,ou=Jenkins,dc=example,dc=com。

2. Jenkins 插件配置

为了实现 Jenkins 与 LDAP 系统集成，需要导航到"插件管理"页面安装 LDAP 插件。

"插件管理"页面如图 2-6 所示。

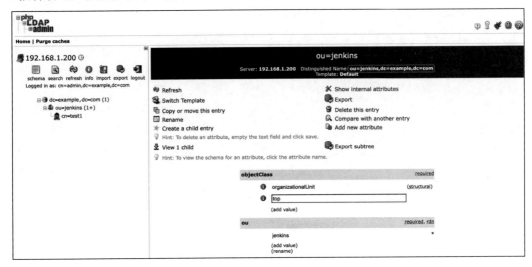

图 2-5　LDAP 系统组织页面

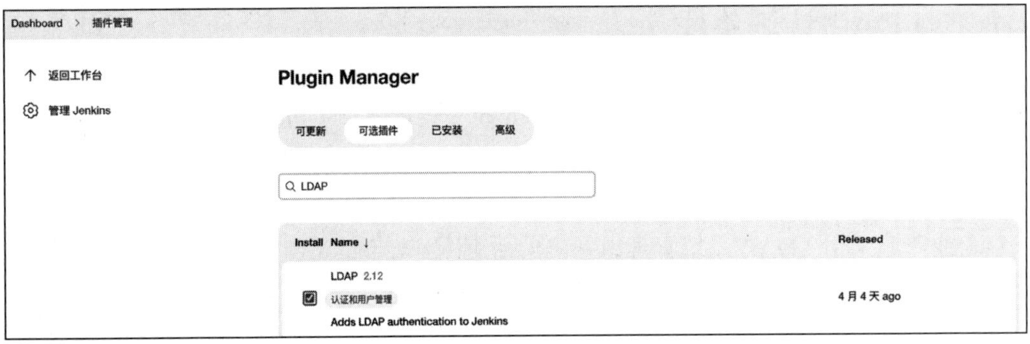

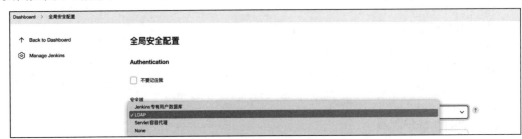

图 2-6　"插件管理"页面

插件安装完成后，导航到"全局安全配置"页面开启 LDAP 认证。"全局安全配置"页面如图 2-7 所示。

图 2-7　"全局安全配置"页面

填写 LDAP 服务器 Server、root DN、User search base、User search filter、Group search base 等信息。设置 LDAP 服务器和组织信息如图 2-8 所示。

图 2-8 设置 LDAP 服务器和组织信息

填写 Manager DN 的管理员账户"cn=admin,dc=example,dc=com"和对应的密码。设置 LDAP 管理员账户如图 2-9 所示。

图 2-9 设置 LDAP 管理员账户

配置好 LDAP 参数后需要重启 Jenkins Server 方可生效，登录时使用的是 LDAP 用户 test1（cn=test1,ou=Jenkins,dc=example,dc=com）。登录成功后，可以在用户列表中看到当前的用户。Jenkins 系统用户列表如图 2-10 所示。

图 2-10　Jenkins 系统用户列表

2.2　权限管理

　　Jenkins 的权限管理是一种安全实践，皆在控制平台上的不同租户对 Jenkins 项目的访问范围。若企业中存在若干项目组使用同一个 Jenkins 来进行 CI/CD，则需要为每个项目组配置权限，减少误触发的情况。

　　在企业中通过创建 Jenkins 项目命名规范标准来识别不同组织的项目。Jenkins 用户与权限图如图 2-11 所示，UserA 属于 devops 团队，具有以 devops 为前缀的 Jenkins 项目的权限；UserB 属于 test 团队，具有以 test 为前缀的 Jenkins 项目的权限。

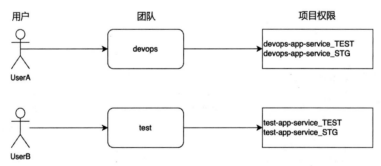

图 2-11　Jenkins 用户与权限图

2.2.1　准备

　　为了后面演示权限管理的效果，笔者准备创建 1 个测试用户 jenkins_user1 和 4 个测试 Jenkins 项目。

　　导航到用户管理页面，添加测试用户 jenkins_user1。添加测试用户后的系统用户列表如图 2-12 所示。

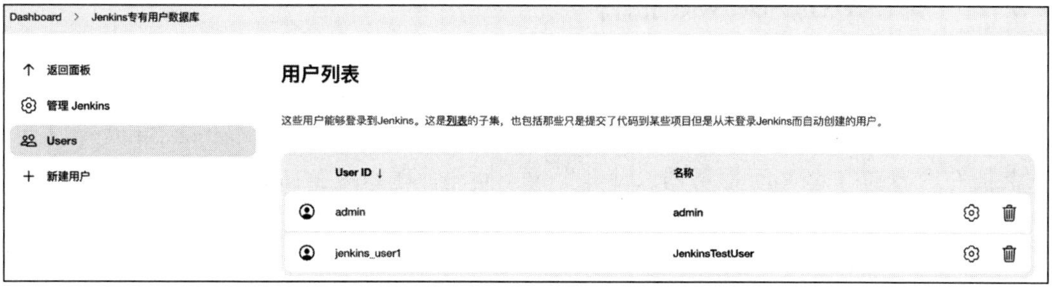

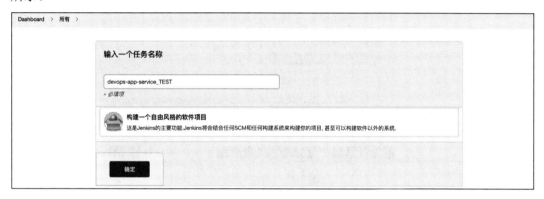

图 2-12　添加测试用户后的系统用户列表

导航到新建任务页面，分别按照同样的步骤创建 4 个测试任务。项目创建页面如图 2-13 所示。

图 2-13　项目创建页面

如图 2-14 所示，Jenkins 项目分别是 devops-app-service_STG、devops-app-service_TEST、test-app-service_STG 和 test-app-service_TEST。

图 2-14　项目页面

2.2.2　安装 Role-based 插件

在 Jenkins 系统中，授权插件应用最广的是 Role-based Authorization Strategy。导航到插件管理页面，单击"可选插件"选项卡，选择安装 Role-based Authorization Strategy 插件。安装插件后需重启 Jenkins 服务器以使其生效。插件管理如图 2-15 所示。

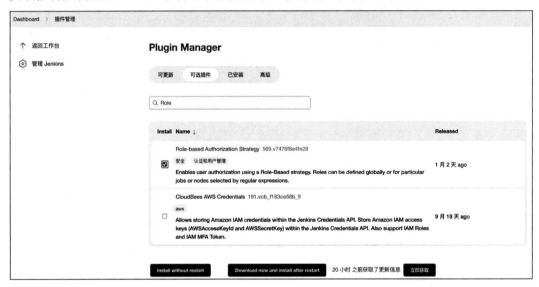

图 2-15　插件管理

插件安装成功后，导航到"全局安全配置"页面，配置授权策略为 Role-Based Strategy。然后单击"保存"按钮完成配置。全局安全配置的配置授权策略如图 2-16 所示。

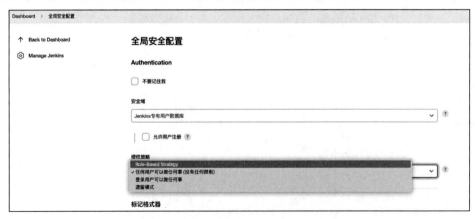

图 2-16　全局安全配置的配置授权策略

完成上面的配置后，再次导航到系统管理页面，可以看到新的选项"Manage and Assign Roles"，如图 2-17 所示。

图 2-17　Jenkins 系统管理页面之管理角色

选择图 2-17 所示的"Manage and Assign Roles"选项后，可以看到"Manage Roles""Assign Roles""角色策略宏"选项。Roles 管理页面如图 2-18 所示。

图 2-18　Roles 管理页面

2.2.3　创建角色

在图 2-18 中选择 Manage Roles 进入管理角色页面，可以看到以下 3 种角色。
- ☑　Global roles：全局权限配置，权限是由 Item Roles 和 Node Roles 组成的集合。

- ☑ Item roles：项目级权限配置，配置项目权限。
- ☑ Node roles：节点及权限配置，配置节点权限（用途很少，本节省略其示例）。

1．创建 Global roles

创建 Global roles 为 devopsdev，并分配开发人员常用的任务构建和配置权限。开发人员的权限在企业中基本上相同，后面所有项目组的开发人员都可以使用此角色。创建 Global roles 页面，如图 2-19 所示。

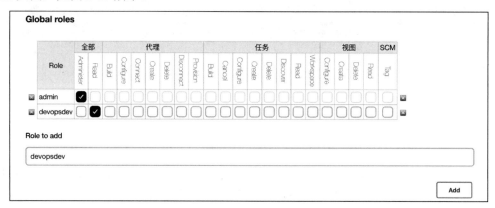

图 2-19　创建 Global roles 页面

2．创建 Item roles

创建 Item roles 为 devops，使用正则表达式设置权限的作用范围。这里匹配的是项目名称以 devops-开头并且以任意字符结尾的项目，将具有项目的 Build、Cancel、Configure、Read 权限，创建 Item roles 页面如图 2-20 所示。

图 2-20　创建 Item roles 页面

2.2.4 授权角色

参考图 2-18，进入 Assign Roles 页面。

1. 授权 Global roles

为 jenkins_user1 用户授权 devopsdev 全局角色，授权 Global roles 页面如图 2-21 所示。

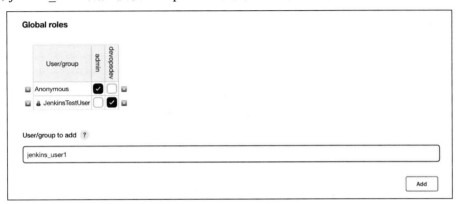

图 2-21　授权 Global roles 页面

2. 授权 Item roles

为 jenkins_user1 用户授权 devops 项目权限。授权 Item roles 页面如图 2-22 所示。

图 2-22　授权 Item roles 页面

2.2.5　测试权限

使用 jenkins_user1 账号登入 Jenkins，如果仅能看到 devops 相关的项目就表示成功实现了权限控制。每个项目组中的用户仅可操作所属项目组内的项目。Jenkins 登录成功后首页如图 2-23 所示。

图 2-23　Jenkins 登录成功后首页

提示：

如果提示"Access Denied"，说明没有给用户分配 Global roles 中的全部 Read 权限。Jenkins 用户无权限报错如图 2-24 所示。

图 2-24　Jenkins 用户无权限报错

2.3　凭据管理

在实施 DevOps 的过程中，以 Jenkins 作为核心引擎，就少不了与其他系统集成和交互。此时，Jenkins 需要使用一些系统的认证信息。例如，Jenkins 系统需要下载代码仓库，就需要使用代码库的账号和密码。本节将讲解 Jenkins 凭据（credentials）的管理。

2.3.1　安装凭据插件

账号密码属于敏感信息，稍有操作不当就会产生泄露风险。因此不能使用明文或硬编码的方式保存。为了解决这些问题，可以使用凭据插件管理用户账号密码信息实现加密存储，而且不会在日志中输出。安装凭据插件页面如图 2-25 所示。

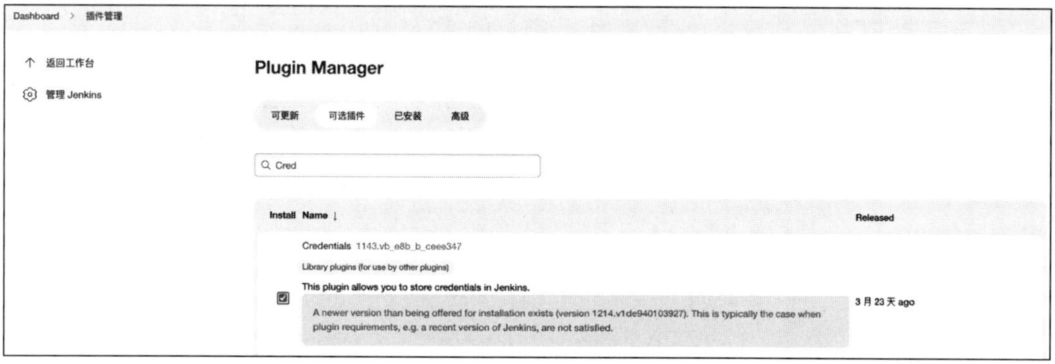

图 2-25　安装凭据插件页面

插件安装成功后，可以在系统管理页面中看到 Manage Credentials 选项。安装凭据后的系统管理页面如图 2-26 所示。

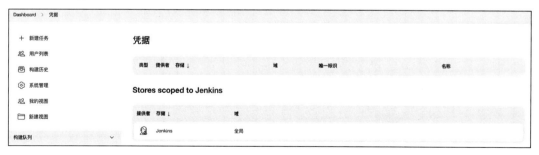

图 2-26　安装凭据后的系统管理页面

选择 Manage Credentials 选项进入"凭据"页面。凭据默认存在全局的域。"凭据"页面如图 2-27 所示。

图 2-27　"凭据"页面

单击"全局"进入全局域，可以看到当前域的所有凭据信息。凭据列表页面如图 2-28 所示。

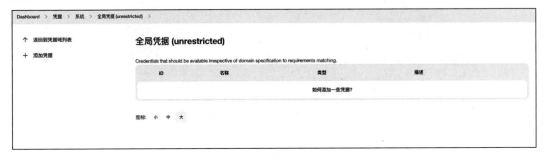

图 2-28　凭据列表页面

2.3.2　创建凭据

单击"添加凭据"按钮，选择凭据"类型"，创建凭据页面如图 2-29 所示。例如：存储 GitLab 仓库的账号和密码，需要填写用户名、密码、描述等信息，然后单击 Create 按钮完成创建。

图 2-29　创建凭据页面

创建好凭据之后，可以看到一串凭据 ID，后续我们在 Jenkins Pipeline 实践时会通过此 ID 对凭据进行引用。"全局凭据（unrestricted）"列表页面如图 2-30 所示。

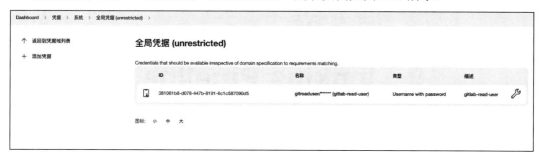

图 2-30　"全局凭据（unrestricted）"列表页面

凭据的类型有很多种，当添加凭据后找不到对应类型时，需要在插件管理页面安装对应类型的凭据插件。

2.4　本章小结

本章主要讲解了持续集成系统 Jenkins 在企业应用中常用的配置实践。为读者展示了 Jenkins 企业级中的用户管理、权限管理、凭据管理。至此，读者已经掌握了 Jenkins 在企业中的配置实践技能，能够根据企业现状制定相应的解决方案。

下一章将讲解持续集成系统 Jenkins 的核心内容——Jenkins Pipeline 实战。笔者将结合自身的工作经验为读者介绍一些企业中常用的 Jenkins 的管理实践。

第 3 章
Jenkins Pipeline 实战

Jenkins Pipeline 使用 Groovy 脚本语言来定义流水线，这些脚本可以访问 Jenkins 管理员配置的插件，并与外部系统（如版本控制系统、云平台等工具系统）集成。通过 Jenkins Pipeline，用户可以在 Jenkins 环境中实现 CI/CD 工作流，并可以通过定义多个阶段和任务来实现自动化流程。本章，笔者将逐一展示 Jenkins Pipeline 的语法与实践。

本章内容速览。
- ☑ 什么是 Pipeline
- ☑ Pipeline 核心语法
- ☑ Pipeline 开发工具
- ☑ 共享库实践
- ☑ 本章小结

3.1 什么是 Pipeline

Jenkins Pipeline（简称 Pipeline）支持在 Jenkins 中实施持续集成和持续交付流水线。采用 Pipeline 可以自动化实现日常重复性运维和测试任务。

3.1.1 Pipeline 简介

Jenkins 的核心是流水线类型项目，实现了流水线即代码（Pipeline as Code）。我们可以将构建、部署、测试等步骤以代码的方式定义，在运行 Pipeline 时 Jenkins 会按照文件中定义的代码顺序执行。下面是一个简单的 Pipeline 代码展示。

```
pipeline {
    agent any
    stages {
```

```
        stage('Checkout') {
            steps {
                git url: 'https://github.com/your_repo/your_project.git'
            }
        }
        stage('Build') {
            steps {
                sh './gradlew build'
            }
        }
        stage('Test') {
            steps {
                sh './gradlew test'
            }
        }
    }
    post {
        always {
            junit 'build/test-results/*.xml'
        }
    }
}
```

在上面的示例中，Pipeline 包含 3 个阶段。第一个阶段，Checkout 使用 git 插件从 GitHub 存储库中检出源代码。第二个阶段，Build 使用./gradlew build 命令构建项目。第三个阶段，Test 使用./gradlew test 命令执行自动化测试。最后，post 部分负责收集测试结果报告。

> **提示：**
> 如果 Pipeline 运行时无法展示阶段视图，则需要安装 StageView 插件。

3.1.2 为什么使用 Pipeline

Jenkins Pipeline 是 Jenkins 持续交付流水线的一种方法，用于实现持续交付的工作流。它允许用户通过代码定义和自动化构建、测试和部署流程。使用 Jenkins Pipeline 可以帮助用户实现更快、更准确、可重复地构建，测试和部署流程，从而提高持续交付速度和交付质量。Pipeline 的特点如下。

- ☑ 自动化：Jenkins Pipeline 可以使用代码定义持续交付流程，从而实现自动化。这可以减少人工干预，降低出错的风险，并提高效率。
- ☑ 易于维护：Jenkins Pipeline 使用代码定义流程，因此易于维护和更新。如果需要更改流程，则只更改代码即可。

- ☑ 易于查看和追踪：Jenkins Pipeline 可以在 Jenkins 中生成图形界面，方便用户查看和追踪流程的状态。
- ☑ 集成：Jenkins Pipeline 可以与外部系统集成，从而实现更强大的持续交付流程。
- ☑ 可扩展性：Jenkins Pipeline 可以使用各种插件扩展功能，从而满足用户的不同需求。

3.1.3 什么是 Jenkinsfile

Jenkinsfile 是一种用于描述 Jenkins Pipeline 的文本文件。它使用 Pipeline 语法定义构建、部署和测试工作流程的步骤。Jenkinsfile 是按需生成的，存储在源代码管理系统（如 Git）中，以便每次代码更改时都可以重复使用。使用 Jenkinsfile 可以确保构建、部署和测试过程的可重复性和可预测性，并且可以在多个项目之间共享和重复使用。版本控制系统的目录结构如图 3-1 所示。

名称	最后提交	最后更新
📁 src/org/devops	更新 src/org/devops/tools.groovy, Jenkin...	刚刚
📁 vars	更新 vars/hello.groovy	刚刚
📄 Jenkinsfile	更新 src/org/devops/tools.groovy, Jenkin...	刚刚
📄 README.md	更新 src/org/devops/tools.groovy, Jenkin...	刚刚

图 3-1 版本控制系统的目录结构

3.2 Pipeline 核心语法

Jenkins 支持两种编写 Pipeline 代码的语法：脚本式语法和声明式语法。脚本式语法使用 Groovy 脚本语言编写，提供了较多的灵活性和控制，但是需要管理员掌握更多相关的代码和规则。声明式语法更加简洁明了，适合于简单的工作流程，并且不需要编写大量的代码。声明式语法易于维护和理解。

选择使用哪种语法取决于具体的需求和个人喜好。如果计划对 Jenkins Pipeline 代码有更高的定制需求，可以使用脚本式语法；如果计划编写简单的 Pipeline 代码，则可以使用声明式语法。接下来将展示 Pipeline 的核心语法。

📖 **提示：**

笔者推荐在声明式语法中嵌入脚本式语法，这样可以结合两种语法的优点。在 steps{}

语句块中添加script{}语句块，即可实现运行脚本式语法。代码如下。

```
pipeline{
   agent any
   stages{
      stage("test"){
         steps{
            script{
                //插入脚本式语法代码
            }
         }
      }
   }
}
```

3.2.1　agent 节点

在 Pipeline 语法中，agent{}语句可以指定 Pipeline 的阶段在哪个机器上运行。通常可以通过名称、标签等方式选择机器，也可以通过以下参数定义其他运行方式。

- ☑ any：在任意节点运行。
- ☑ none：不指定节点运行。
- ☑ lable：在指定特定名称或者标签的节点运行。

在任意节点运行的参考代码如下。

```
pipeline {
   agent any
}
```

在指定名称或者标签的节点运行的参考代码如下。

```
pipeline {
   agent { label "label Name" }
}
```

在 Pipeline 中不指定节点，也可以在阶段中指定节点，参考代码如下。

```
pipeline {
   agent none
   stages{
      stage('Build'){
         agent { label "build" }          //在阶段中指定运行节点
         steps {
            echo "building..."
```

```
            }
        }
    }
}
```

3.2.2 stages 阶段

一条 Jenkins Pipeline 通常具有许多阶段。在 Pipeline 的语法中，使用 stages{}语句块定义所有的阶段，其中每个阶段使用 stage{} 语句块进行定义。每个 stage{} 语句块中包含一个 steps{}语句块，该语句块用于指定要运行的步骤。代码如下。

```
pipeline {
    agent { label "build" 1}
    stages{
        stage('Build'){
            steps {
                echo "building..."          //打印消息
            }
        }
    }
}
```

在上面的示例中，Pipeline 运行在具有 build 标签或者名称为 build 的节点上。该 Pipeline 包含一个阶段 Build。Pipeline 阶段数量取决于实际的项目情况，没有固定一说。我们可以将构建打包、单元测试、自动化测试、自动部署分别通过阶段进行定义。

3.2.3 post 构建后操作

在 Pipeline 的运行结束后，通常我们会快速反馈给触发人或者邮件组等人员。使用 post{}语句块可以根据 Pipeline 不同的状态进行不同的操作。常用的状态参数如下。

- ☑ always：不判断状态，总是运行。
- ☑ success：Pipeline 成功后运行。
- ☑ failure：Pipeline 失败后运行。
- ☑ aborted：Pipeline 被取消后运行。

代码如下。

```
pipeline {
    agent none
    stages{
```

```
        stage('Build'){
            agent { label "build" }
            steps {
                echo "building..."
            }
        }
    }
    post {
        always{
            script{
                println("pipeline 结束后，经常做的事情")
            }
        }
        success{
            script{
                println("pipeline 成功后，要做的事情")
            }
        }
        failure{
            script{
                println("pipeline 失败后，要做的事情")
            }
        }
        aborted{
            script{
                println("pipeline 取消后，要做的事情")
            }
        }
    }
}
```

3.2.4　environment 环境变量

在开发复杂的 Pipeline 过程中，需要用到一些参数，这些参数可以通过环境变量来定义。在 Pipeline 的阶段中可以获取环境变量。环境变量在 Pipeline 中也分为全局变量和局部变量。

全局变量在 pipeline{} 中定义，会在所有阶段中生效。局部变量在 stage{} 中定义，仅在当前阶段生效。代码如下。

```
pipeline {
    agent { label "build" }
    environment {
```

```
            version = "1.1.1"
            envType = "dev"
    }
    stages{
        stage('Build'){
            environment {
                version = "1.1.2"
            }
            steps {
                echo "${version}"   //引用变量
                echo "${envType}"
            }
        }
    }
}
```

> **提示：**
> 当相同的变量名称在全局和局部都有定义时，局部变量的优先级要高于全局变量，即局部变量生效。

3.2.5　options 运行选项

Pipeline 的运行可以通过参数控制，如 Pipeline 失败时可以重试的次数、Pipeline 运行超时的时间等。使用 options{}语句块可以定义 Pipeline 的运行选项参数。

- ☑ buildDiscarder(logRotator(numToKeepStr: '1'))：设置保存最近的记录。
- ☑ disableConcurrentBuilds()：禁止 Pipeline 并行构建。
- ☑ skipDefaultCheckout()：跳过默认的代码检出阶段。
- ☑ timeout(time: 1, unit: 'HOURS')：Pipeline 超时时间。
- ☑ retry(3)：Pipeline 的重试次数。
- ☑ timestamps()：Pipeline 日志中显示时间戳。

options{}语句块可以在 pipeline{}语句块中定义，也可以在 stage{}语句块中定义。代码如下。

```
pipeline {
    options {
        disableConcurrentBuilds()
        skipDefaultCheckout()
        timeout(time: 1, unit: 'HOURS')
    }
```

```
stages {
    stage("build"){
        options {
            timeout(time: 5, unit: 'MINUTES')
            retry(3)
            timestamps()
        }
    }
}
```

> **提示：**
> 当使用 timestamps() 运行参数时，需要提前安装好 Timestamper 插件。

3.2.6　parameters 参数

当 Pipeline 触发前需要让用户选择或者自定义一些参数时，可以使用 parameters{}语句块。该语句块可以在代码中定义参数，代码初始化之后会将这些参数添加到 UI 页面，当用户触发 Pipeline 时可以看到这些参数。

常用的参数类型如下。

- ☑ string：字符串参数。
- ☑ choice：选项参数。

下面的示例定义了两个参数：字符串参数和选项参数。代码如下。

```
pipeline {
    agent any
    parameters {
        string name: 'VERSION', defaultValue: '1.1.1', description: ''
        choice choices: ['dev', 'test', 'uat'], description: 'env names', name: 'ENVNAME'
    }
    stages {
        stage("Build"){
            steps {
                echo "${params.VERSION}"
                echo "${params.ENVNAME}"
            }
        }
    }
}
```

在 Jenkins 系统中创建一个 Pipeline 作业，并复制上面代码立即构建运行。运行成功后再次打开此作业会发现，页面上多了两个我们在 Pipeline 中定义的参数（见图 3-2）。这样就便于用户选择参数触发 Pipeline，操作更加灵活。

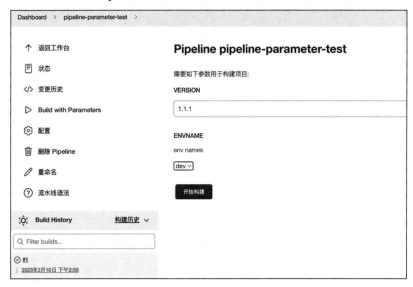

图 3-2　Pipeline 参数化构建

提示：

Pipeline 的参数化构建可以使用户交互更加灵活。这些参数可以通过作业设置页面在图形化页面中添加，也可以通过 parameters{}语句块定义。需要注意的是，如果通过代码定义参数，必须在执行一次以后才能在作业页面显示出来。

3.2.7　triggers 触发器

企业中存在自动运行 Pipeline 的场景，例如，每天晚上 11 点进行一些测试作业，第二天上班后就可以看到测试报告，可见触发器提高了工作效率。triggers{}语句块可以配置以下触发方式。

- ☑ cron()：定时触发，使用计划任务语法定义。
- ☑ pollSCM()：被动检测源代码管理系统存在变更后触发。
- ☑ upstream()：由上游作业触发，可以根据作业状态判断。

代码如下：

```
pipeline {
```

```
    agent any
    triggers {
        cron('H */7 * * 1-5')    //定时触发
        //upstream(upstreamProjects: 'job1,job2',    //上游作业成功后触发
        //    threshold: hudson.model.Result.SUCCESS)
        //pollSCM('H */7 * * 1-5')    //被动检测源代码管理系统触发
    }
    stages {
        stage('build') {
            steps {
                echo 'Hello World'
            }
        }
    }
}
```

3.2.8 input 交互

使用 Jenkins Pipeline 可以将工作流程一键触发并自动运行，但有时在一些场景下需要设置卡点，经过用户确认后再进行后续的工作。例如，我们在进行配置持续部署流水线时，对应多个环境，不会直接将版本同时发布到所有环境，更多的是先发布测试环境，再发布预生产和生产环境，以保证不会对生产环境造成服务不可用。

通过 input{}语句块可以在阶段运行前，在 Pipeline 运行页面弹出一个选项框和用户交互。代码如下。

```
pipeline {
    agent any
    stages {
        stage('Deploy') {
            input {
                message "是否继续发布并选择发布环境"
                parameters {
                    string(name: 'ENVTYPE', defaultValue: 'DEV', description: 'env type..[DEV/STAG/PROD]')
                }
            }
            steps {
                echo "Deploy to ${ENVTYPE}, doing..."
            }
        }
    }
}
```

代码的运行效果如图 3-3 所示。

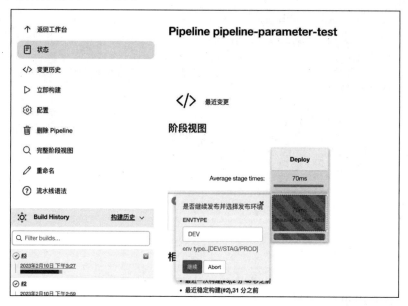

图 3-3　input 交付效果

3.2.9　when 阶段运行控制

在 DevOps 实施过程中，Pipeline 通常具有代码扫描、单元测试等阶段，这些阶段会比较耗时，当紧急变更或者在特殊情况下需要跳过这些阶段时，可以使用 when{}语句块对阶段运行控制。

when{}语句块可以根据条件进行判断，当条件成立时运行对应的阶段。常用的类型如下。

- ☑　branch：根据分支名称判断。
- ☑　environment：根据环境变量值判断。
- ☑　triggeredBy：根据触发类型判断。
- ☑　expression：根据表达式结果判断。

当 DEPLOY_ENV 的值为 prd 时，执行 Deploy 步骤。

```
pipeline {
    agent any
    environment{
        DEPLOY_ENV = "dev"
    }
```

```
stages {
    stage('Build') {
        steps {
            echo 'build...'
        }
    }
    stage('Deploy') {
        when {
            environment name: 'DEPLOY_ENV', value: 'prd'
        }
        steps {
            echo 'Deploying...'
        }
    }
}
```

代码运行效果如图 3-4 所示。

3.2.10 parallel 并行运行

并行运行多个阶段的场景有很多，例如，当产品需要进行自动化测试时可以同时在不同的操作系统环境进行测试。parallel{}语句块用于定义并行运行的阶段。

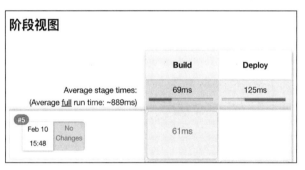

图 3-4　阶段跳过效果图

在不同的运行节点上并行运行多个阶段。

```
pipeline {
    agent any
    stages {
        stage('Parallel Stage') {
            failFast true    //当其中一个阶段失败，立即失败
            parallel {
                stage('windows') {
                    agent {
                        label "master"
                    }
                    steps {
                        echo "windows"
                    }
                }
```

```
        stage('linux') {
            agent {
                label "build"
            }
            steps {
                echo "linux"
            }
        }
    }
}
```

3.3　Pipeline 开发工具

开发 Pipeline 除了掌握语法之外，利用好开发工具会事半功倍。选择任意 Pipeline 类型的作业，单击"流水线语法"即可进入 Pipeline 开发工具页面，如图 3-5 所示。

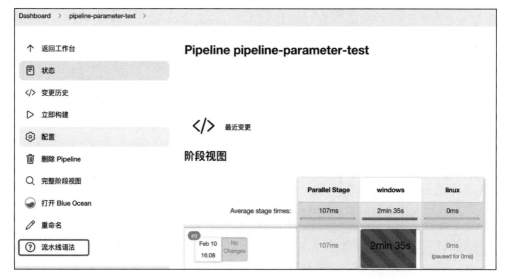

图 3-5　流水线语法菜单

1. 片段生成器

流水线代码片段生成器可以找到每个插件以及 Jenkins 内置方法的使用方法。使用片段生成器可以根据个人需要生成代码，有些方法来源于插件，则需要先安装相关的插件才能找到对应的方法，如图 3-6 所示。

图 3-6　片段生成器

2．声明式语法生成器

声明式语法生成器（declarative directive generator）可以用于生成对应的语法代码，如图 3-7 所示。

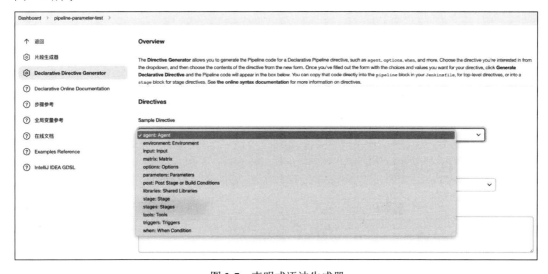

图 3-7　声明式语法生成器

3. 全局变量参考

全局变量是 Jenkins 系统内置的变量。用户可以根据需求引用这些变量，例如，获取项目构建 ID、项目构建时间、项目的 URL 等信息，如图 3-8 所示。

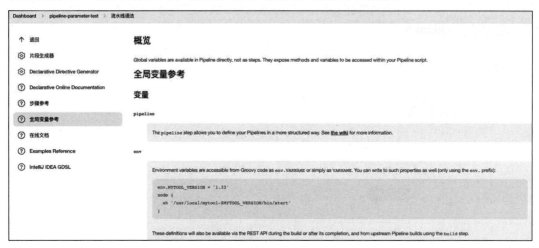

图 3-8　Jenkins 全局变量参考

4. 流水线回放

流水线回放功能可以方便用户在不修改项目设置时使用原 Pipeline 代码进行调试，如图 3-9 所示。

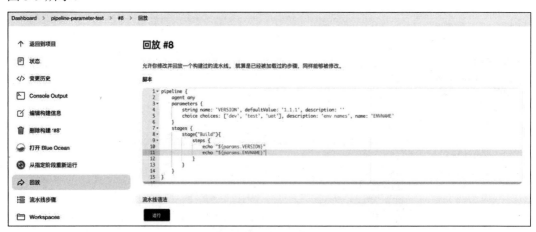

图 3-9　流水线回放

3.4 共享库实践

共享库（Jenkins Shared Library）是一个可以在 Jenkins Pipeline 中共享和重复使用的代码库。它允许用户存储和维护 Pipeline 代码的公共部分，并将其作为其他 Pipeline 的依赖项进行加载。

3.4.1 创建共享库

共享库通常存储在版本控制系统（如 Git）中，并可以在多个 Jenkins 实例上共享。首先，需要在 Git 存储库中创建一个共享库以存储公共的代码。然后，可以使用 Groovy 脚本编写代码。我们采用 GitLab 作为版本控制系统，共享库目录结构如图 3-10 所示。

图 3-10　共享库目录结构

共享库目录结构类似于 Java 工程结构，每个目录都有其对应的作用。

☑　vars：此目录存放 Pipeline 中使用的函数。

- ☑ src：此目录存放共享代码，如类和辅助函数。
- ☑ resources：此目录包含可在 Pipeline 中使用的资源，如配置文件和 Shell 脚本。
- ☑ docs：此目录包含帮助文档和示例，以帮助用户理解如何使用。

提示：

vars 目录和 src 目录有些区别，vars 目录定义的方法可以全局使用且无须导入。src 目录中定义的方法需要先导入，才能在 Pipeline 中正常使用。

3.4.2 编写共享库代码

首先，在 vars 目录中创建一个名为 hello.groovy 的 Groovy 文件，然后编写一个 GetFiles() 函数，此函数并没有实际功能，只是通过 println() 函数打印输出日志。代码如下。

```
//hello.groovy
def GetFiles(){
   println("Get files")
}
```

我们在 src 目录下创建一个递归目录 org/devops/，然后新建 Demo.groovy，定义 PrintMsg() 函数，该函数接收一个参数并打印输出日志。代码如下。

```
package org.devops

def PrintMsg(value){
   println(value)
}
```

编写好代码后，提交到版本控制系统。

提示：

现在编写的代码函数并没有实际的功能，仅用于帮助大家快速入门。后续章节会提供具有实际功能的函数。

3.4.3 修改全局设置

通过 3.4.1 节和 3.4.2 节我们完成了共享库的创建并编写了相关的测试代码，接下来我们需要让 Jenkins 知道这个共享库，即安装 Pipeline: Groovy Libraries 插件并配置此共享库。导航到 Jenkins 系统设置，然后找到 Global Pipeline Libraries 选项，单击"新增"按钮新增共享库，如图 3-11 所示。

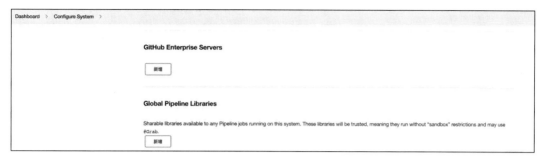

图 3-11　新增共享库

接下来填写共享库的名称和版本信息，共享库名称可以自定义，版本对应的是 Git 中分支的名称。此处注意，根据后来新的版本控制系统将默认 Git 分支从 master 变成 main，按照实际情况填写，此处用到的是 main 分支，所以这里的默认版本填写的是 main。共享库名称和版本配置如图 3-12 所示。

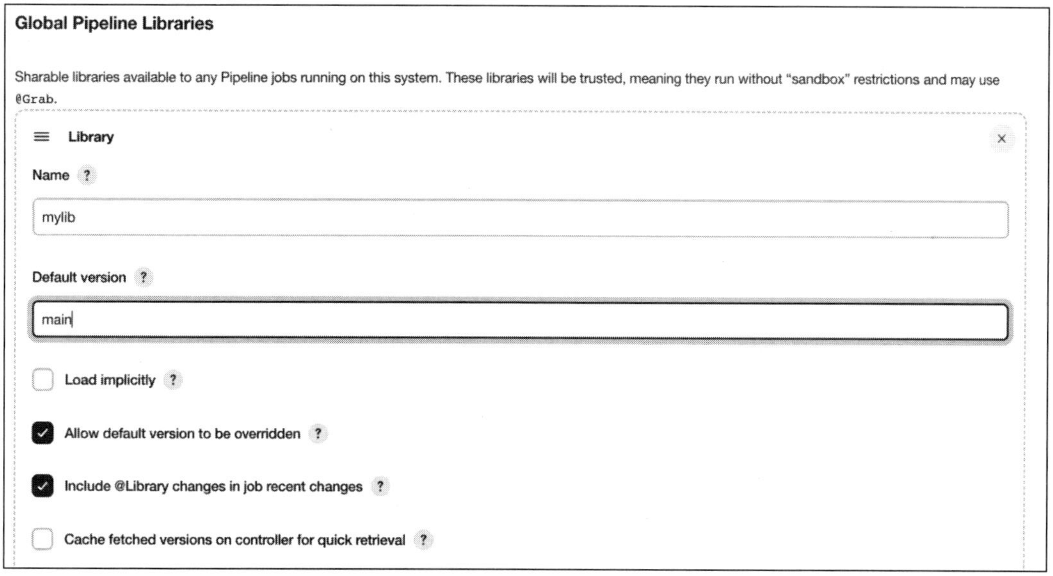

图 3-12　共享库名称和版本配置

由于共享库是 Git 存储库，我们需要填写具体的项目链接信息。在 Source Code Management 下拉列表框中选择 Git。在"项目仓库"文本框中填写项目仓库地址。项目的类型如果是公开的，项目可以不选择凭据；如果是私有的，则需要配置凭据。如果凭据不存在，需要单击"添加"按钮进行添加。共享库项目配置参数如图 3-13 所示。

按照图 3-13 中参数填写完成后，单击"保存"按钮即可应用配置。

图 3-13　共享库项目配置参数

3.4.4　加载共享库

接下来我们通过编写 Jenkinsfile 定义 Pipeline 并加载上述步骤创建的共享库。为了便于管理，我们也将在共享库 Git 仓库中的根目录下创建一个 Jenkinsfile 文件。在实际环境中，可以选择将 Jenkinsfile 和应用项目代码放到一起，或解耦单独存放到一个 Git 仓库。编写 Jenkinsfile 加载共享库，代码如下。

```
//加载共享库
@Library("mylib@main") _

//导入src中的类和函数
def mydemo = new org.devops.Demo()

pipeline{
    agent any
    stages{
        stage("test"){
            steps{
                script{
```

```
            //pipeline 中调用 Demo.groovy 中的方法
            mydemo.PrintMsg("hello 这是 src 中的方法输出")

            //pipeline 中调用 vars/Hello.groovy 中的方法
            hello.GetFiles()
          }
        }
      }
    }
}
```

将此段代码复制到任意 Pipeline 运行。运行结果如图 3-14 所示。

```
[Pipeline] Start of Pipeline
[Pipeline] node
Running on Jenkins in /var/lib/jenkins/workspace/pipeline-parameter-test
[Pipeline] {
[Pipeline] stage
[Pipeline] { (test)
[Pipeline] script
[Pipeline] {
[Pipeline] echo
hello 这是src中的方法输出
[Pipeline] echo
Get files
[Pipeline] }
[Pipeline] // script
[Pipeline] }
[Pipeline] // stage
[Pipeline] }
[Pipeline] // node
[Pipeline] End of Pipeline
Finished: SUCCESS
```

图 3-14　Pipeline 加载共享库成功

以上，我们完成了一个简单的共享库开发，并且加载成功。在开发 Pipeline 过程中，可以在共享库存储使用频繁的构建步骤，以便在多个管道中重复使用它们。这样可以确保管道的一致性，并降低维护管道的难度。

3.5　本章小结

本章主要讲解了 Jenkins Pipeline 实践，为读者展示了 Pipeline 入门、Pineline 核心语法、Pipeline 开发工具、共享库实践。至此，读者已经掌握了 Jenkins Pipeline 的开发技能，能够根据企业现状开发相应的 Jenkins Pipeline。共享库是企业级 Pipeline 的一种最佳实践，后续的实践章节也会以共享库的方式讲解。

第 4 章
Jenkins Pipeline 进阶

本章主要讲解关于 Jenkins 和其他 DevOps 工具链系统的集成准备知识。例如，与版本控制系统集成、获取代码提交分支并进行构建等。为了使 Pipeline 更加灵活，可以借助 Groovy 编程来进行数据处理。Jenkins Pipeline 原生支持 Groovy 语言，所以可以轻松实现功能扩展。接下来笔者将根据自己的集成经验为读者展示以上实践。

本章内容速览。
- ☑ Groovy 编程语法
- ☑ Jenkins 触发器
- ☑ 常用的 DSL 语句
- ☑ 本章小结

4.1 Groovy 编程语法

Groovy 是一种面向对象的、动态的、基于 JVM 的编程语言。它的语法与 Java 语法非常相似，但同时又拥有灵活的动态特性，并且支持闭包和高阶函数。Groovy 用于实现构建、测试和部署自动化，特别是在 Jenkins 和 Gradle 等 DevOps 工具中。它提供了更简洁、更灵活的语法。本节将讲解 Jenkins Pipeline 中常用的 Groovy 语法内容（并不会深入讲解所有 Groovy 的特性）。

4.1.1 数据类型

Groovy 是一种基于 JVM 的面向对象的脚本语言。Groovy 支持类型自动推导，因此不需要指定变量的类型。Groovy 中的常见数据类型如下。
- ☑ 字符串类型：用单引号或双引号括起来的字符串。
- ☑ 布尔类型：true 和 false。

☑ 集合类型：List、Map 和 Set。

在 Pipeline 中进行字符串类型数据处理，代码如下。

```
//String
//在 Pipeline 外部定义变量（全局生效）
name = "devops"
pipeline {
    agent any
    stages{
        stage("run"){
            steps{
                script{
                    //打印
                    println(name)

                    //在 Pipeline 内定义变量（该阶段生效）
                    job_name = "devops04-app-service_CI"

                    //使用 split()方法进行字符串分割
                    //["devops05", "app", "service_CI"]
                    bu_name = job_name.split('-')[0]
                    println(bu_name)   //devops05

                    //使用 contains()方法判断字符串是否包含特定字符
                    println(job_name.contains("CI"))

                    //使用 size()或 length()方法计算字符串的长度
                    println("size: ${job_name.size()}")
                    println("length: ${job_name.length()}")

                    //使用 endsWith()方法判断字符串是否以特定字符结尾
                    println("enswith CI: ${job_name.endsWith('CI')}")
                }
            }
        }
    }
}
```

以上示例使用了以下字符串处理函数。

☑ split()：函数具有一个分隔符参数，对字符串按照分隔符进行分割。

☑ contains()：函数接收一个字符参数，用于判断接收的字符串是否存在。

☑ size()：用于计算字符串的长度。

☑ length()：用于计算字符串的长度。

☑ endsWith()：接收一个字符参数，用于判断是否以某个字符串结尾。

在 Pipeline 中进行 List 类型数据处理，代码如下。

```
//List
//在 Pipeline 外部定义 List
tools = ["gitlab", "jenkins", "maven", "sonar"]

pipeline {
    agent any
    stages{
        stage("run"){
            steps{
                script{
                    //打印
                    println(tools)

                    //向 List 添加元素
                    println(tools + "k8s")
                    println(tools << "ansible")
                    tools.add("maven")
                    println(tools)

                    //从 List 中移除元素
                    println(tools - "maven")
                    println(tools)

                    //用 contains() 判断元素是否存在于 List 中
                    println(tools.contains("jenkins"))

                    //用 size() 计算 List 中元素的个数
                    println(tools.size())

                    //通过索引获取元素
                    println(tools[0])      //0 为第一个元素
                    println(tools[-1])     //-1 为最后一个元素
                }
            }
        }
    }
}
```

以上示例使用了以下 List 类型数据处理函数。

☑ add()：添加元素。

☑ contains()：判断元素是否存在。

- ☑ size()：计算 List 的长度。
- ☑ index：通过索引定位元素。

在 Pipeline 中进行 Map 类型数据处理，代码如下。

```
//Map

//在 Pipeline 外部定义 Map
user_info = ["id": 100, "name": "jenkins"]
pipeline {
    agent any
    stages{
        stage("run"){
            steps{
                script{
                    //打印
                    println(user_info)

                    //获取 Map 中 key 对应的值
                    println(user_info["name"])
                    println(user_info["id"])

                    //为 Map 赋值
                    user_info["name"] = "jenkinsX"
                    println(user_info)

                    //判断 Map 中是否包含某个 key
                    println(user_info.containsKey("name"))

                    //判断 Map 中是否包含某个 value
                    println(user_info.containsValue(100))

                    //获取 Map 中的 key 集合
                    println(user_info.keySet())

                    //删除 Map 中的 key
                    user_info.remove("name")
                    println(user_info)
                }
            }
        }
    }
}
```

以上是在 Pipeline 中经常使用的数据类型和数据处理方法。在开发一些可复用、可扩展的 Pipeline 时会用到这些。

4.1.2 控制语句

Groovy 控制语句用于控制代码执行流。对于开发高效、稳健的应用程序是非常重要的。常用的 Groovy 控制语句如下。

- ☑ if 语句：用于判断布尔表达式的真假，并在布尔表达式为真时执行一段代码。
- ☑ switch 语句：用于在多个条件之间进行选择，并在匹配到一个条件时执行相应的代码。
- ☑ for 语句：用于在特定范围内迭代一系列值。
- ☑ while 语句：用于在特定条件为真时重复执行一段代码。

1．在 Pipeline 中使用 if 语句

代码如下。

```
//if
//在 Pipeline 外部定义变量
branchName = "dev"

pipeline {
    agent any
    stages{
        stage("run"){
            steps{
                script {
                    if ( branchName == "dev"){
                        println("deploy to dev....")
                    } else if (branchName == "master"){
                        println("deploy to stag...")
                    } else {
                        println("error...")
                    }
                }
            }
        }
    }
}
```

以上示例使用 if 语句判断 branchName 变量的值。如果值为 dev，则打印"deploy to dev..."；如果条件不成立，则继续判断值是否为 master，如果是，则打印"deploy to stag..."；如果不是，则进入 else 语句块打印"error..."。

2. 在 Pipeline 中使用 switch 语句

代码如下。

```
//switch
branchName = "dev"

pipeline {
    agent any
    stages{
        stage("run"){
            steps{
                script {
                    switch(branchName) {
                        case "dev":
                            println("deploy to dev...")
                            break
                        case "master":
                            println("deploy to stag...")
                            break
                        default:
                            println("error...")
                            break
                    }
                }
            }
        }
    }
}
```

以上示例使用 switch 语句判断 branchName 变量，每个匹配通过 case 定义。如果值为 dev，则打印 "deploy to dev..."，break 跳出 switch 语句；如果值为 master，则打印 "deploy to stag..." 并跳出 switch 语句；如果上面的 case 条件都不匹配，则进入 default 语句块打印 "error..." 并跳出 switch 语句。

> **提示：**
>
> 在 Groovy 的 switch 语句语法中，每个 case 语句结尾都添加 break 来跳出 switch 语句，如果未添加，即使匹配了 case 语句，还是会继续往下匹配。

3. 在 Pipeline 中使用 for 循环语句

代码如下。

```
//for
```

```
users = [
        ["name": "devops", "role": "dev"],
        ["name": "devops1", "role": "admin"],
        ["name": "devops2", "role": "ops"],
        ["name": "devops3", "role": "test"]
    ]

pipeline {
    agent any
    stages{
        stage("run"){
            steps{
                script {
                    user_names = []
                    for (i in users){
                        println(i["name"])
                        user_names << i["name"]
                    }
                    println(user_names)// [devops,devops1,devops2,devops3]
                }
            }
        }
    }
}
```

以上示例使用 for 循环语句遍历 users 列表,遍历 users 中的每个元素并将 name 对应的值添加到新的 List,即 user_names 中。

4. 在 Pipeline 中使用 while 循环语句

代码如下。

```
//while

sleeps = true

pipeline {
    agent any
    stages{
        stage("run"){
            steps{
                script {
                    while(sleeps){
                        println("sleep...")
                    }
```

```
                }
            }
        }
    }
}
```

以上示例使用 while 语句判断 sleeps 的值，条件成立会一直打印"sleep..."直到条件不成立。

4.1.3 异常处理

Groovy 支持标准的异常处理机制，语法和 Java 语言类似。异常处理的语法如下。

```
//try...catch...finally
pipeline{
    agent any
    stages{
        stage("run"){
            steps{
                script{
                    try {
                        println(a)
                    } catch(Exception e){
                        println(e)
                    } finally {
                        println("always...")
                    }
                }
            }
        }
    }
}
```

以上示例中，程序块 try 内的代码可能抛出异常，因为变量 a 未定义；catch 块内的代码用于处理异常；finally 块内的代码不管是否抛出异常都会执行。

4.1.4 函数

Groovy 函数是一个命名的、独立的代码块，可以在代码中多次调用。需要重复执行的代码逻辑可以封装成函数，避免代码块重复。

在 Pipeline 中使用函数，代码如下。

```
//function
users = [
        ["id": 1, "name": "jenkins1"],
        ["id": 2, "name": "jenkins2"],
        ["id": 3, "name": "jenkins3"],
    ]

pipeline{
    agent any
    stages{
        stage("run"){
            steps{
                script {
                    //调用函数
                    name = GetUserNameByID(1)
                    println(name)   //jenkins1
                }
            }
        }
    }
}
def GetUserNameByID(id){
    for (i in users){
        if (i["id"] == id){
            return i["name"]
        }
    }
    return "null"
}
```

以上示例，def 关键字定义函数名为 GetUserNameByID，此函数带有一个参数 id；函数体内的 for 循环遍历 users 列表。如果 id 与参数 id 一致，则返回用户的 name；否则返回 null，然后在 pipeline{}内部可以通过函数名称和参数调用此函数。

4.2　Jenkins 触发器

笔者在大量的 Jenkins Pipeline 实践中发现，处理任务有时需要借助其他系统触发 Jenkins，然后传递一些参数，Pipeline 要拿到这部分参数然后进行处理。例如，开发人员在 GitLab 版本控制系统中提交了代码，此时需要 GitLab 将此次事件的数据发送给 Jenkins，而 Jenkins Pipeline 需要解析这些数据并获取分支参数，然后进行构建。

4.2.1 安装触发器

通常使用 Generic Webhook Trigger 插件来配置 Jenkins 触发器。它允许使用 Webhook 将外部系统与 Jenkins 集成。通过配置 Webhook，在特定事件发生时，外部系统可以触发 Jenkins 构建。例如，可以在源代码管理系统中触发构建、在问题跟踪系统中触发构建等。

在"Dashboard/插件管理"中安装 Generic Webhook Trigger 插件，如图 4-1 所示。

图 4-1　安装 Generic Webhook Trigger 插件

然后进入任意一个 Pipeline 类型的项目中开启此触发器，如图 4-2 所示。

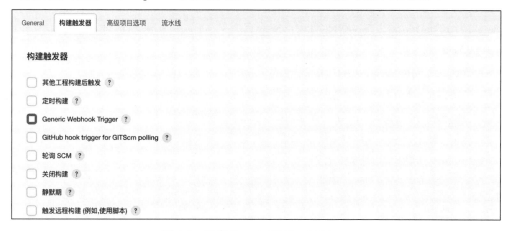

图 4-2　开启 Generic Webhook Trigger

4.2.2 配置触发器

我们在项目中开启 Generic Webhook Trigger 触发器之后，可以看到一些参数，如图 4-3

所示。

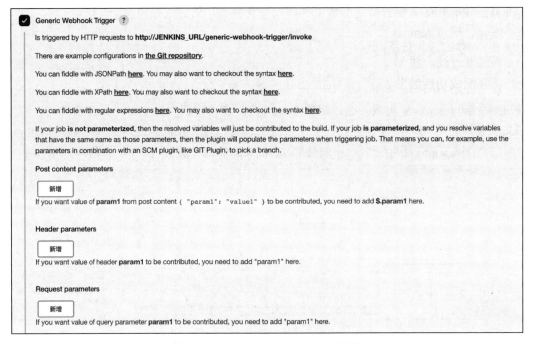

图 4-3　Generic Webhook Trigger 配置

接收外部其他系统请求的 Webhook URL 是 http://JENKINS_URL/generic-webhook-trigger/invoke。调用的时候要把 JENKINS_URL 换成真实的 Jenkins 服务器地址，有端口就加上端口，是域名就写域名。可以从 Webhook 中过滤的参数如下。

- ☑　Post content parameters：配置过滤请求传递的 body 部分数据。
- ☑　Header parameters：配置过滤请求中的 header 部分数据。
- ☑　Request parameters：配置过滤请求 URL 中传递的参数。

为了防止触发冲突，可以给项目配置一个触发的 Token 参数，使用 Token 匹配一个或者一组 Pipeline 作业，如图 4-4 所示。

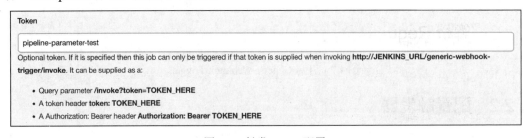

图 4-4　触发 Token 配置

第 4 章 Jenkins Pipeline 进阶

> **提示：**
> 为多个项目配置相同的 Token 会被同时触发，避免触发冲突，Token 的值要具有唯一性。例如，使用 Jenkins 作业名称作为 Token 值等。

当需要过滤一些 Webhook 触发时，可以通过 Optional filter 使用正则表达式和值进行判断，匹配成功后触发 Pipeline 任务运行，如图 4-5 所示。

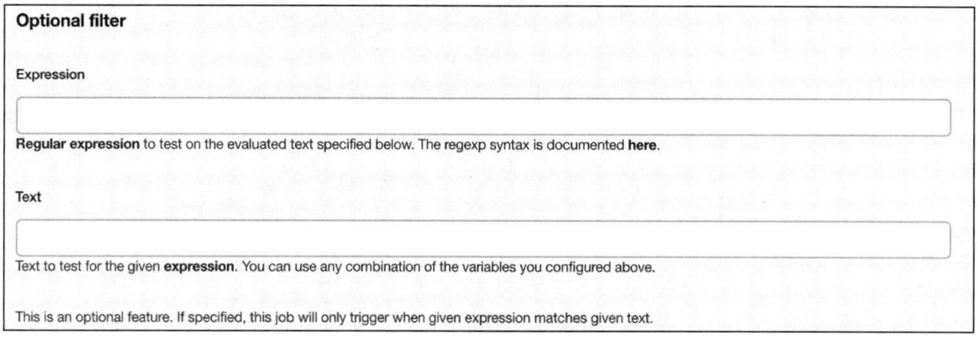

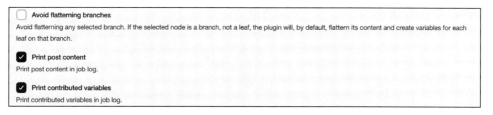

图 4-5　过滤选项配置

> **提示：**
> 如果配置了过滤选项，而 Pipeline 未如期触发，则应检查此部分的过滤选项配置。可能是由于正则表达式和值不匹配导致未触发成功。

为了方便调试此触发器，建议选择以下两个选项来打印调试的日志到 Pipeline，如图 4-6 所示。

图 4-6　开启调试日志

4.2.3　解析 Request 参数

下面通过 curl 命令来模拟请求测试，准备传递参数 version 和 username。curl 命令如下。

```
curl "http://192.168.1.200:8080/generic-webhook-trigger/invoke?token=pipeline-parameter-test&version=1.1.1&username=jenkins"
```

配置 Generic Webhook Trigger 接收 version、username 这两个参数，如图 4-7 所示。

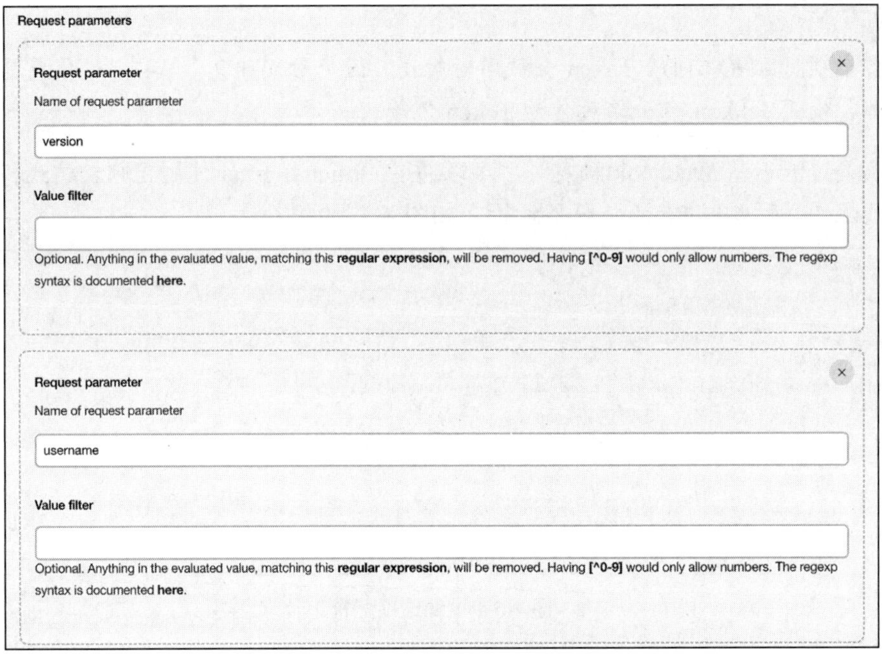

图 4-7　Request parameters 配置

> **提示：**
>
> 图 4-7 中的参数名称要和实际请求传递的参数名称一致，否则会出现空值的情况。特殊情况下也可以在 Value filter 中通过编写正则表达式对值进行再次过滤。

编写一条 Pipeline 来读取 version、username 变量，代码如下。

```
println("${username}")
println("${version}")

pipeline{
    agent any
    stages{
        stage("test"){
            steps{
                script{
                    echo "${username}"
                    echo "${version}"
                }
            }
        }
    }
}
```

保存上述配置，在终端中运行 curl 命令测试触发 Jenkins 作业，命令如下。

```
curl http://192.168.1.200:8080/generic-webhook-trigger/invoke?token=
pipeline-parameter-test&version=1.1.1&username=jenkins
#响应
{
    "jobs":{
        "pipeline-parameter-test":{
            "regexpFilterExpression":"",
            "triggered":true,
            "resolvedVariables":{
                "username":"jenkins",
                "username_0":"jenkins",
                "version":"1.1.1",
                "version_0":"1.1.1"
            },
            "regexpFilterText":"",
            "id":40,
            "url":"queue/item/40/"
        }
    },
    "message":"Triggered jobs."
}
```

打开 Jenkins Pipeline，然后查看输出的日志，如图 4-8 所示。

```
[Pipeline] echo
jenkins
[Pipeline] echo
1.1.1
[Pipeline] node
Running on Jenkins in /var/lib/jenkins/workspace/pipeline-parameter-test
[Pipeline] {
[Pipeline] stage
[Pipeline] { (test)
[Pipeline] script
[Pipeline] {
[Pipeline] echo
jenkins
[Pipeline] echo
1.1.1
[Pipeline] }
[Pipeline] // script
[Pipeline] }
[Pipeline] // stage
[Pipeline] }
[Pipeline] // node
[Pipeline] End of Pipeline
Finished: SUCCESS
```

图 4-8 Pipeline 日志输出

以上就是模拟请求传递参数之后 Jenkins Pipeline 读取参数的整个过程。

4.2.4 解析 Header 参数

通过以下 curl 命令来模拟请求测试。传递两个 Header 参数：header_name 和 header_id。

```
curl --location --request GET 'http://192.168.1.200:8080/generic-webhook-
trigger/invoke?token=pipeline-parameter-test' \
--header 'header_name: jenkins' \
--header 'header_id: 100'
```

配置 Generic Webhook Trigger 接收 header_name、header_id 这两个参数，如图 4-9 所示。

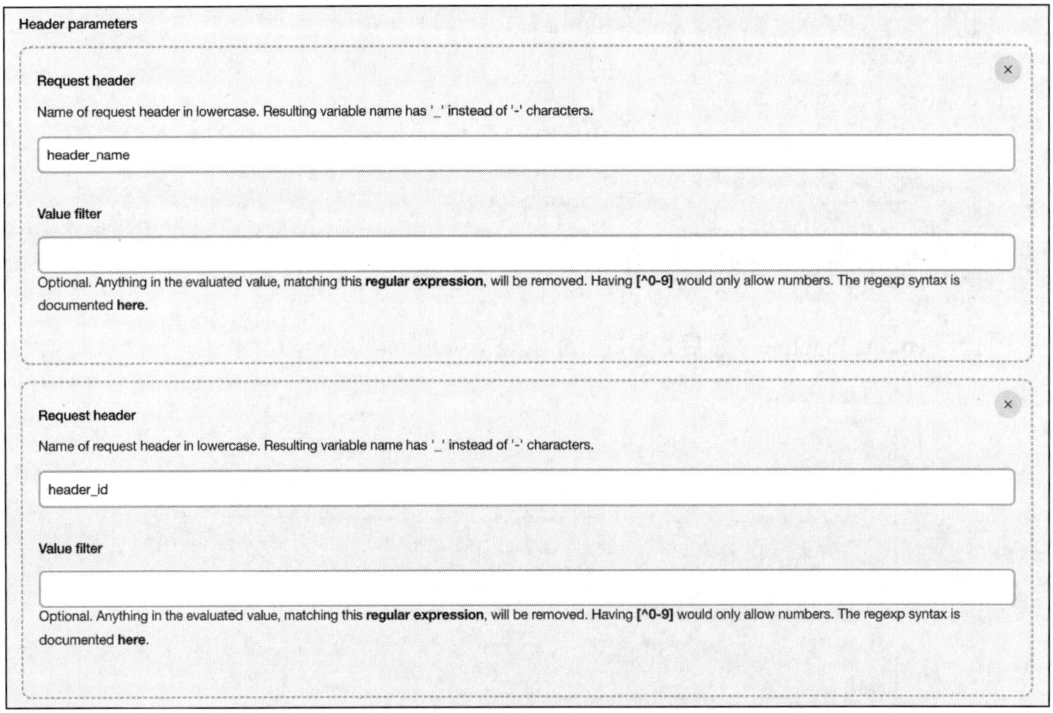

图 4-9 Header parameters 配置

提示：

图 4-9 中的参数名称要和实际请求传递的参数名称一致，否则会出现空值的情况。特殊情况下也可以在 Value filter 中通过编写正则表达式对值进行再次过滤。

编写一条 Pipeline 来读取 header_name、header_id 变量，代码如下。

```
println("${header_name}")
println("${header_id}")

pipeline{
    agent any
    stages{
        stage("test"){
            steps{
                script{
                    echo "${header_name}"
                    echo "${header_id}"
                }
            }
        }
    }
}
```

保存上述配置,然后在终端运行 curl 命令测试触发 Jenkins 作业。命令如下。

```
curl --location --request GET 'http://192.168.1.200:8080/generic-webhook-trigger/invoke?token=pipeline-parameter-test' \
--header 'header_name: jenkins' \
--header 'header_id: 100'
#响应
{
    "jobs":{
        "pipeline-parameter-test":{
            "regexpFilterExpression":"",
            "triggered":true,
            "resolvedVariables":{
                "":"",
                "header_id":"100",
                "header_id_0":"100",
                "header_name":"jenkins",
                "header_name_0":"jenkins"
            },
            "regexpFilterText":"",
            "id":44,
            "url":"queue/item/44/"
        }
    },
    "message":"Triggered jobs."
}
```

打开 Jenkins Pipeline，查看输出的日志，如图 4-10 所示。

```
[Pipeline] echo
jenkins
[Pipeline] echo
100
[Pipeline] node
Running on Jenkins in /var/lib/jenkins/workspace/pipeline-parameter-test
[Pipeline] {
[Pipeline] stage
[Pipeline] { (test)
[Pipeline] script
[Pipeline] {
[Pipeline] echo
jenkins
[Pipeline] echo
100
[Pipeline] }
[Pipeline] // script
[Pipeline] }
[Pipeline] // stage
[Pipeline] }
[Pipeline] // node
[Pipeline] End of Pipeline
Finished: SUCCESS
```

图 4-10　Pipeline 日志输出

以上就是模拟请求传递参数后 Jenkins Pipeline 读取参数的整个过程。

4.2.5　解析 Post 参数

通过 curl 命令来模拟请求测试，传递 JSON 参数。命令如下。

```
curl --location --request POST 'http://192.168.1.200:8080/generic-webhook-trigger/invoke?token=pipeline-parameter-test' \
--header 'Content-Type: application/json' \
--data-raw '{
    "name": "jenkins",
    "id": "123",
    "group1": {
        "name": "jenkins",
        "id" : "001",
        "age": "40"
    }
}'
```

配置 Generic Webhook Trigger 接收这些 JSON 参数，如图 4-11 所示。

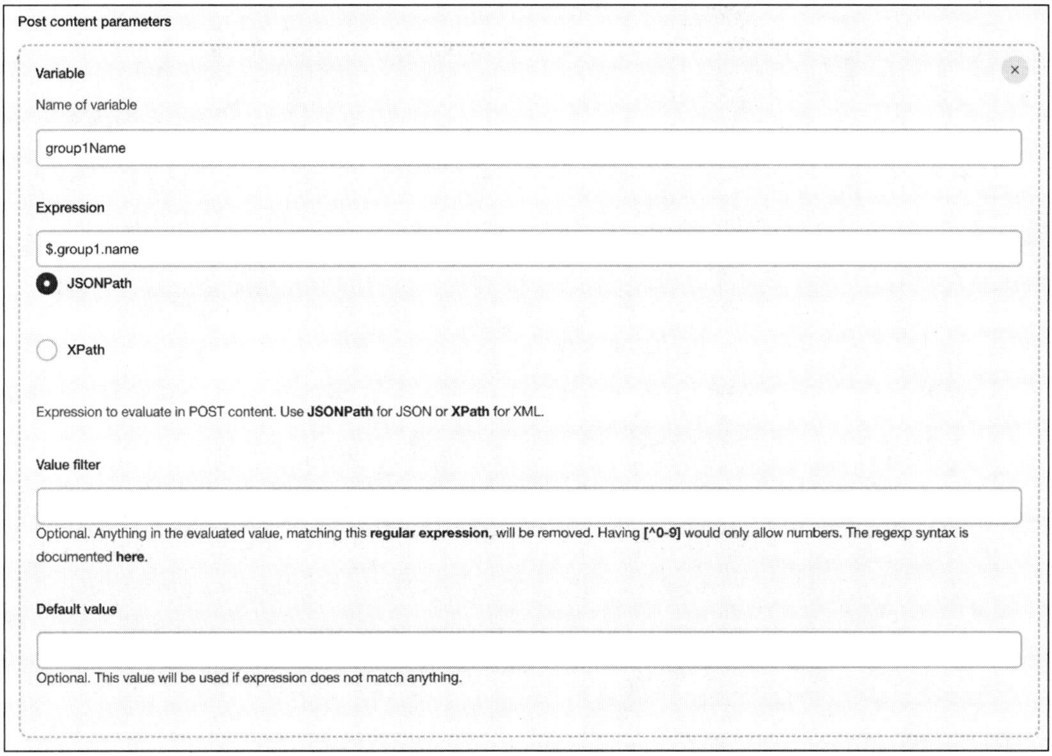

图 4-11　Post content parameters 配置

> **提示：**
>
> 图 4-11 中的参数名称可以自定义，而 Expression 是用来获取数据的。最后，把 Expression 获取的数据赋值给变量 group1Name，Pipeline 中使用变量 group1Name 来获取该参数的值。

另外，Expression 表达式的语法支持 JSONPath 和 XPath 两种类型。图 4-11 所示为使用 JSONPath 格式获取的 JSON 数据，其中 $ 是根节点。

编写一条 Pipeline 读取 header_name、header_id 变量，代码如下。

```
println("${group1Name}")

pipeline{
    agent any
    stages{
        stage("test"){
            steps{
```

```
                script{
                    echo "${group1Name}"
                }
            }
        }
    }
}
```

保存上述配置，然后在终端中运行 curl 命令测试触发 Jenkins 作业。命令如下。

```
curl --location --request POST 'http://192.168.1.200:8080/generic-webhook-trigger/invoke?token=pipeline-parameter-test' \
--header 'Content-Type: application/json' \
--data-raw '{
    "name": "jenkins",
    "id": "123",
    "group1": {
        "name": "jenkins",
        "id" : "001",
        "age": "40"
    }
}'
#响应
{
    "jobs":{
        "pipeline-parameter-test":{
            "regexpFilterExpression":"",
            "triggered":true,
            "resolvedVariables":{
                "group1Name":"jenkins"
            },
            "regexpFilterText":"",
            "id":47,
            "url":"queue/item/47/"
        }
    },
    "message":"Triggered jobs."
}
```

打开 Jenkins Pipeline，查看输出的日志，如图 4-12 所示。

以上就是模拟请求传递参数，使用 Jenkins Pipeline 读取参数的整个过程。

```
[Pipeline] echo
jenkins
[Pipeline] node
Running on Jenkins in /var/lib/jenkins/workspace/pipeline-parameter-test
[Pipeline] {
[Pipeline] stage
[Pipeline] { (test)
[Pipeline] script
[Pipeline] {
[Pipeline] echo
jenkins
[Pipeline] }
[Pipeline] // script
[Pipeline] }
[Pipeline] // stage
[Pipeline] }
[Pipeline] // node
[Pipeline] End of Pipeline
Finished: SUCCESS
```

图 4-12　Pipeline 日志输出

4.3　常用的 DSL 语句

DSL 指的是领域特定语言，Jenkins Pipeline 中有很多基于 Groovy 封装的 DSL，集成 DSL 便于对 Pipeline 进行扩展。

4.3.1　获取当前触发用户

在做邮件通知时，我们可以将 Pipeline 运行状态发送给触发人。安装 build user vars 插件可以支持我们通过 Pipeline 获取触发者的用户名称，安装 build user vars 插件如图 4-13 所示。

图 4-13　安装 build user vars 插件

在 Pipeline 中使用 BuildUser 方法，代码如下。

```
pipeline{
    agent any
    stages{
        stage("test"){
            steps{
                script{
                    wrap([$class: 'BuildUser']){
                        echo "full name is $BUILD_USER"
                        echo "user id is $BUILD_USER_ID"
                        echo "user email is $BUILD_USER_EMAIL"
                    }
                }
            }
        }
    }
}
```

运行 Pipeline，日志输出如图 4-14 所示。

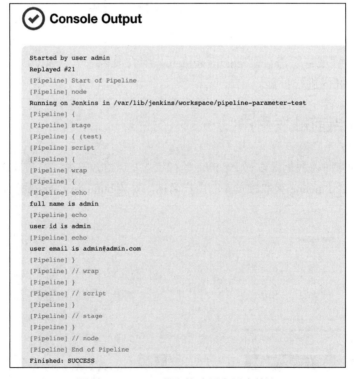

图 4-14　Jenkins 获取构建用户日志输出

4.3.2 JSON 数据解析

如果外部系统传递给 Pipeline 的是 JSON 格式的数据，就需要格式化处理才能获取其中的值。需要安装 Pipeline Utility Steps 插件来使用其中的数据处理方法，如图 4-15 所示。

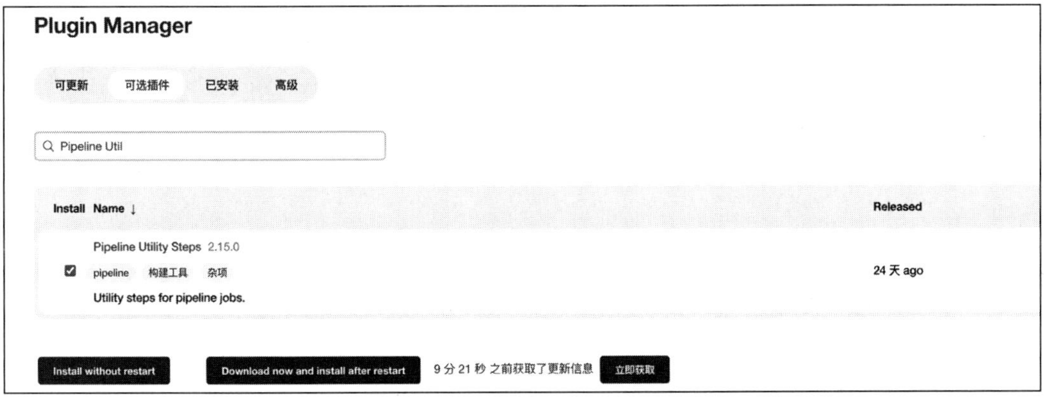

图 4-15 安装 Pipeline Utility Steps 插件

在 Pipeline 中使用 readJSON 方法，代码如下。

```
jsonData = """
{
  "name":"devops",
  "id" : 100
}
"""

pipeline{
    agent any
    stages {
        stage("test"){
            steps{
                script{
                    data = readJSON text: "${jsonData}"
                    println(data.name)
                }
            }
        }
    }
}
```

运行 Pipeline，日志输出如图 4-16 所示。

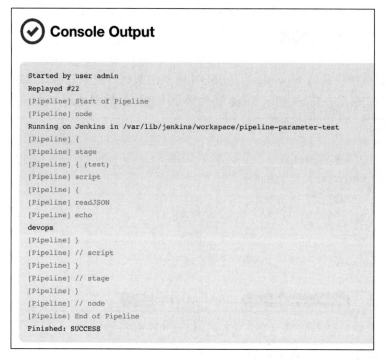

图 4-16　Jenkins 解析 JSON 数据日志输出

4.3.3　在 Pipeline 中使用凭据

我们可以将与其他系统集成所需要的验证用户和密码存储到 Jenkins 凭据中，也可以在 Pipeline 的代码中引用凭据。

在 Pipeline 中使用凭据方法，代码如下。

```
pipeline{
    agent any
    stages{
        stage("test"){
            steps{
                script{
                    withCredentials([usernamePassword(credentialsId: '381061b7-d077-447b-8191-6c1c587090d5',
                        passwordVariable: 'PASSWORD',
                        usernameVariable: 'USERNAME')]) {

                        sh "curl -u ${USERNAME}:${PASSWORD} http://127.0,0,1"
                    }
```

```
            }
        }
    }
}
```

4.3.4 自定义构建 ID 和描述

Jenkins 中默认的构建 ID 是迭代的数字，例如默认第一次构建作业时，构建 ID 就是数字 1。可以通过 currentBuild.displayName 修改构建 ID。为了便于查看当前作业的简短描述，可以通过 currentBuild.description 定义。

在 Pipeline 中自定义构建 ID 和描述的方法，代码如下。

```
pipeline{
    agent any
    stages{
        stage("test"){
            steps{
                script{
                    currentBuild.displayName = "dev"
                    currentBuild.description = "Hello "
                }
            }
        }
    }
}
```

Pipeline 运行后的效果如图 4-17 所示。

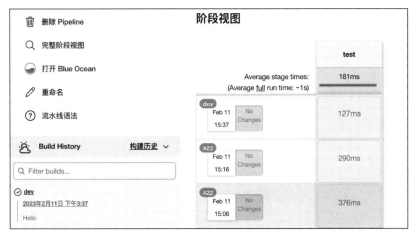

图 4-17　Jenkins 自定义构建 ID 和描述

4.4 本章小结

本章内容是 Pipeline 的进阶部分，通过讲解 Groovy 编程语法皆在使读者掌握 Pipeline 开发的核心技能并能够自由、灵活地通过 Groovy 语言扩展 Pipeline。Jenkins 触发器便于后期与其他 DevOps 工具链系统集成，读者可以复制此实践与其他系统集成。常用的 DSL 方法便于开发和调试 Jenkins Pipeline。

第 5 章 项目代码管理

GitLab 是一个全面的 DevOps 平台，包含版本控制、CI/CD、容器编排、监控等多种功能。它提供了 GitLab CE（社区版）和 GitLab EE（企业版）两个版本。GitLab CE 版适用于个人用户，提供版本控制和 CI/CD 等基础功能；GitLab EE 版适用于企业级用户，提供高级功能。无论个体开发者还是企业级用户，GitLab 提供的一站式开发解决方案，让开发、测试、部署与监控都能够高效有序地进行。本章，笔者将介绍 GitLab 版本控制系统入门和提交流水线实践。

本章内容速览。
- ☑ GitLab 系统入门
- ☑ GitLab 工作流
- ☑ 提交流水线实践
- ☑ 项目构建工具
- ☑ 本章小结

5.1 GitLab 系统入门

GitLab 是一个用于项目管理、仓库管理、源代码管理、持续集成、持续交付和部署的开源的 DevOps 平台，是在企业中应用最广的开源版本控制系统。GitLab 可以帮助团队缩短产品生命周期并提高生产力，从而为客户创造价值。

5.1.1 GitLab 概述

GitLab 是基于 Web 的 Git 存储库，提供免费的开放和私有类型的存储库、问题跟踪功能。同时，GitLab 是一个完整的 DevOps 平台，通过它，专业人员能够执行项目中的所有任务——从项目规划和源代码管理到监控和安全。此外，GitLab 还允许团队协作并构建更好的软件。

使用 GitLab 的主要好处是，允许所有团队成员在项目的每个阶段进行协作。GitLab 提供从计划到创建的跟踪，以帮助开发人员自动化整个 DevOps 生命周期并获得最佳结果。越来越多的开发人员开始使用 GitLab，因为它具有广泛的功能和代码可用性。

5.1.2　GitLab 安装部署

为避免容器删除后数据丢失，启动 GitLab 服务时建议使用 Docker 容器，并创建需要持久化的目录。同时，还需在本地磁盘中创建 GitLab 需要使用的配置目录、日志目录等，确保 GitLab 启动并正常运行。具体而言，需要创建 config、logs、data 等目录，将其挂载到 Docker 容器中，以完成 GitLab 服务的启动及数据的持久化。命令如下。

```
mkdir -p /data/devops/gitlab/{config, logs, data}
chmod +x -R /data/devops/gitlab
```

在启动容器时，通过 Docker 命令行工具中的 -v 参数，将本地目录与 GitLab 容器内的数据目录挂载起来，这样可以将 GitLab 的数据持久化到本地磁盘中，防止容器删除后数据丢失。命令如下。

```
docker run -itd --name gitlab \
-p 443:443 \
-p 80:80 \
--restart always \
--hostname 192.168.1.200 \
-v /data/devops/gitlab/config:/etc/gitlab \
-v /data/devops/gitlab/logs:/var/log/gitlab \
-v /data/devops/gitlab/data:/var/opt/gitlab \
gitlab/gitlab-ce:15.0.3-ce.0
```

容器启动后，可以通过命令 docker logs -f gitlab 查看日志。通过日志可以获取登入系统的账号和密码存储的文件路径。日志如下。

```
Running handlers complete
Cinc Client failed. 285 resources updated in 01 minutes 03 seconds

Notes:
Default admin account has been configured with following details:
Username: root
Password: You didn't opt-in to print initial root password to STDOUT.
Password stored to /etc/gitlab/initial_root_password. This file will be
cleaned up in first
reconfigure run after 24 hours.
```

进入容器并获取密码。访问 http://本机 IP:80 可以进入 GitLab 系统登录页面，如图 5-1 所示。

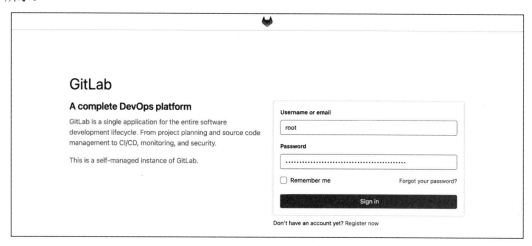

图 5-1　GitLab 系统登录页面

提示：

第一次登入系统后可以选择修改密码。在 GitLab 中登录自己的账号，进入个人资料页面，单击页面右上角的用户头像，选择 Settings 进入用户设置页面。单击页面左侧的 Account，进入账户设置页面。在 Password 页面输入当前密码和新密码，并确认新密码。单击 Save password 按钮保存修改，如图 5-2 所示。

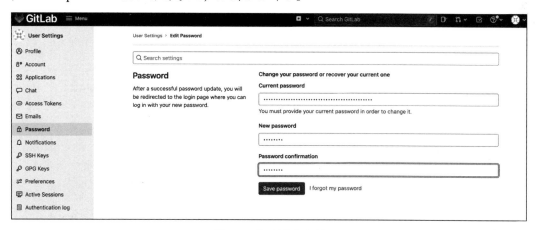

图 5-2　修改用户密码

如果修改成功，GitLab 会弹出提示框，显示"Your account was updated successfully."提示信息。

5.2 GitLab 工作流

GitLab 是一个分布式版本控制系统，可以实现代码的存储、版本控制和协作。它使用 Git 作为代码仓库，支持团队多人协作和开发，保证代码的完整性。

5.2.1 创建项目组和项目

在 GitLab 上面，可以使用项目组对同一组织或者类型的项目进行分类，用项目组来管理项目，用项目来存储实际的代码。

1. 创建项目组

项目组（Group）是 GitLab 中管理项目的主要方式。在项目组下可以创建多个项目，并定义自己的成员和权限。项目组通常是由一个团队或部门的人一起使用的，他们可以共享代码并协同处理任务。

选择 Menu→Groups→Create group 创建项目组，如图 5-3 所示。

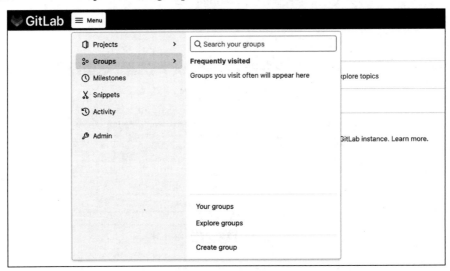

图 5-3　创建 Group 页面

如果存在其他版本控制系统，如 GitHub 等，可以使用 Import group 导入其他系统中的项目组。这里选择 Create group 创建项目组，如图 5-4 所示。

第 5 章　项目代码管理

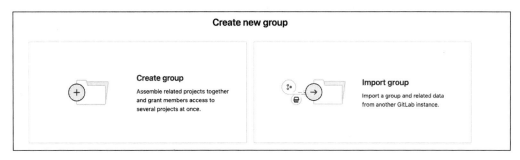

图 5-4　创建项目组的方式

输入项目组的名称。项目组类型分为 Public（公开）、Private（私有）、Internal（内部访问），这里我们选择 Private 类型项目组，单击 Create group 按钮完成创建，如图 5-5 所示。

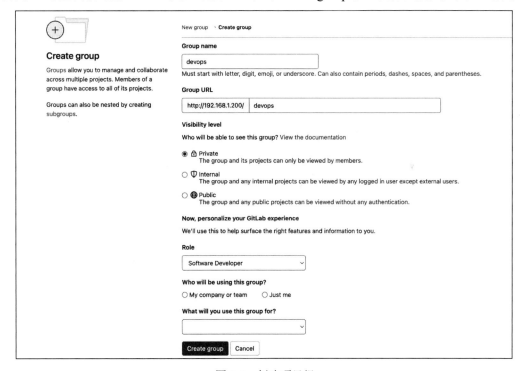

图 5-5　创建项目组

完成项目组的创建后，可以看到当前项目组中还没有项目，如图 5-6 所示。

2．创建项目

项目（project）是 GitLab 中的每个代码库，它们与项目组相关联。每个项目都有一个 Git 存储库用于存储原始的应用代码和其他文件。

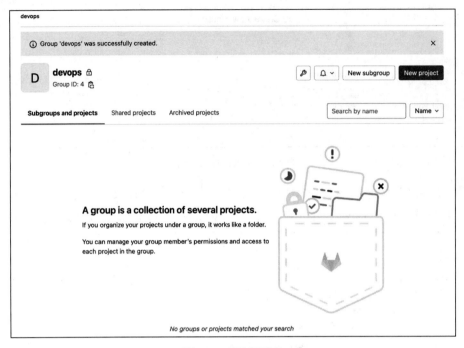

图 5-6　项目组页面

在图 5-6 所示的项目组页面，单击右上角的 New project 按钮来创建项目。GitLab 支持项目的多种类型，可以创建空白项目或者通过 Git 地址从其他系统导入项目。这里我们创建一个空白的项目，选择 Create blank project 类型，如图 5-7 所示。

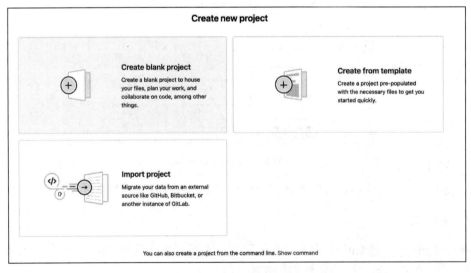

图 5-7　创建 GitLab 项目

第 5 章 项目代码管理

填写项目名称、地址、描述信息、仓库类型，然后单击 Create project 按钮完成项目创建，如图 5-8 所示。

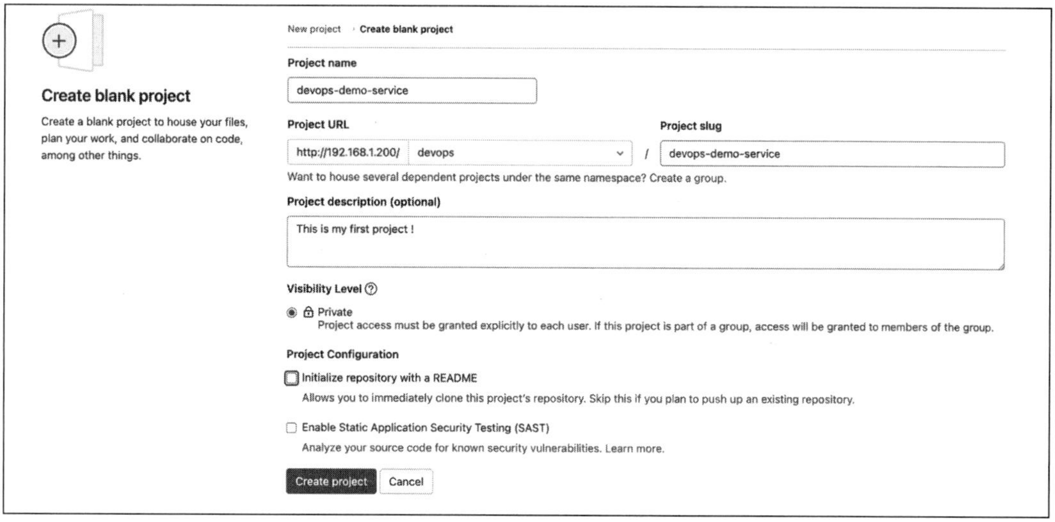

图 5-8　填写项目参数

至此，项目创建成功，如图 5-9 所示。

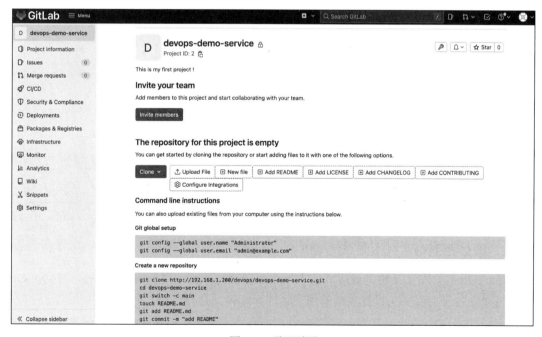

图 5-9　项目页面

5.2.2 生成和提交项目代码

在 GitLab 上面已经创建了一个空的仓库。现在可以创建工程代码并上传到 GitLab 系统中进行版本控制。

1. 生成项目代码

项目技术栈中不同代码的结构和初始化工具也不太一样,这里我们以 Java 语言项目 Spring Boot 为例。在初始化 Spring Boot 应用程序时,可以使用 spring initializr 站点完成。它是一个 Web 应用程序,可以帮助用户快速设置和生成 Spring Boot 项目的构建和配置。

访问链接 https://start.spring.io/ 进入初始化页面,填写项目参数后,单击 GENERATE 按钮初始化生成一个测试项目。初始化参数如图 5-10 所示。

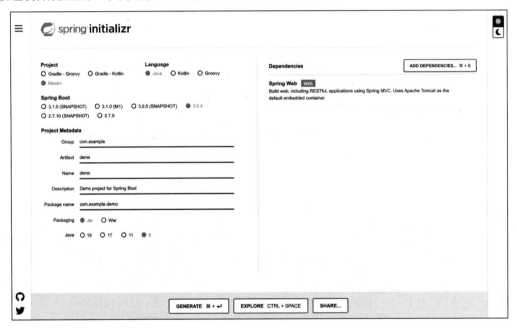

图 5-10 初始化参数

2. 提交项目代码

通过 Git 客户端工具将上面生成的本地代码提交到远程的存储库中。首先,在执行 Git 客户端命令之前,可以先设置当前的用户名称和邮箱。命令如下:

```
git config --global user.name "devops"
git config --global user.email "devops@dev.com"
```

进入本地项目代码目录，然后将代码提交到远程。命令如下。

```
git init
git remote add origin http://192.168.1.200/devops/demo-hello-service.git
git add .
git commit -m "Initial commit"
git push -u origin master
```

下面验证代码是否已经成功地提交到远程存储库，代码库目录如图 5-11 所示。

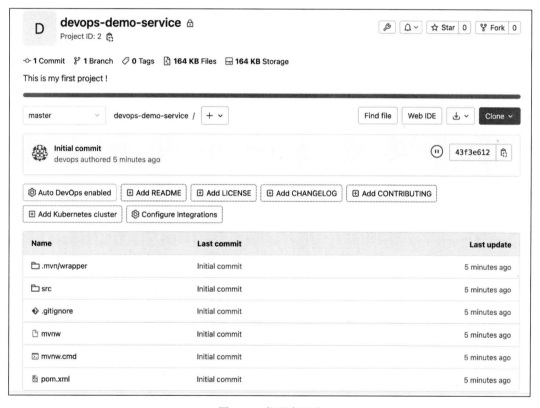

图 5-11　代码库目录

5.2.3　分支开发策略

Git 分支开发策略是一种项目管理和开发流程规范，它规定了 Git 仓库中分支的创建、合并和删除方法，以及团队如何协作开发项目。通过合理使用 Git 分支策略，可以避免不同开发者之间的代码冲突，保证代码的稳定性和版本控制的准确性。常用的 Git 分支开发策略有主干分支开发策略和特性分支开发策略。

1. 主干分支开发策略

基于一个主分支（通常是 main 分支），在此分支上进行所有的开发工作，包括新特性的开发、漏洞修复，并通过主干分支进行应用构建和发布，如图 5-12 所示。

图 5-12　主干分支开发策略

2. 特性分支开发策略

基于主干分支，针对不同的功能或特性，创建独立的分支进行开发。分支的命名通常是基于功能或特性的描述，如 feature/login。每个分支只关注特定的功能或特性，开发完成后，将分支合并到主干分支中，如图 5-13 所示。

图 5-13　特性分支开发策略

本次我们使用特性分支开发策略，并在此分支策略基础上增加了版本分支。也就是说，为避免造成主干分支代码污染，在特性分支开发完成后建立一个版本分支，使特性分支与其合并，最终通过版本分支进行构建和发布，完成后再合并到主干分支。特性分支开发和版本分支开发策略如图 5-14 所示。

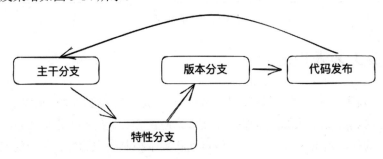

图 5-14　特性分支开发和版本分支开发策略

5.3 提交流水线实践

提交流水线是指在代码提交阶段自动触发 Jenkins Pipeline 任务来对提交分支代码进行编译构建、单元测试、代码扫描等健康检测。使用提交流水线可以轻松实现持续集成，确保每次提交都能够被及时验证并给出反馈。通过提交流水线还可以记录每个步骤的执行情况和结果，为开发和运维人员提供可视化的审计和日志信息。

本节，笔者将把企业中用到的 GitLab 版本控制系统与持续集成系统 Jenkins 的集成实践展示给大家，如图 5-15 所示。

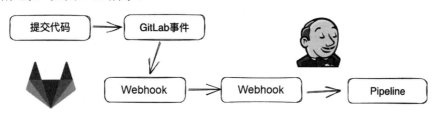

图 5-15 GitLab 与 Jenkins 集成

5.3.1 Jenkins 配置

Jenkins Pipeline 需要被 GitLab 触发。使用 Generic Webhook 插件为 Jenkins Pipeline 开启 Webhook 特性。关于 Generic Webhook Trigger 的用法可以参考 4.2 节。

1. 创建 Pipeline

创建一个新的 Jenkins Pipeline，名称为 devops-demo-service，类型选择"流水线"，如图 5-16 所示。

2. 开启触发器

开启 Generic Webhook Trigger 触发器，添加一个 Post connect parameters 用于捕获 GitLab 传递的数据，如分支名称、项目地址等。这里定义的变量名称为 WebHookData，触发器配置如图 5-17 所示。

上述配置将 GitLab 传递的数据存储到 WebHookData 变量中，在 Jenkins Pipeline 中可以通过变量的方式解析变量的值。

配置 Jenkins Pipeline 的触发 Token。为了保证 Token 的唯一性，这里采用当前项目名称，如图 5-18 所示。

图 5-16　新建 Jenkins Pipeline

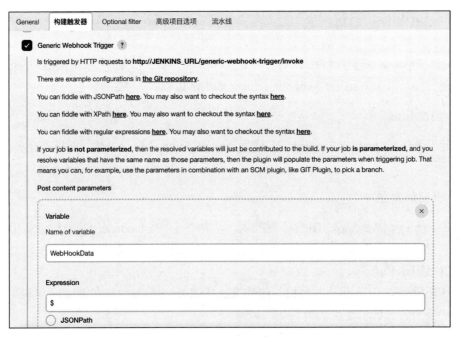

图 5-17　触发器配置

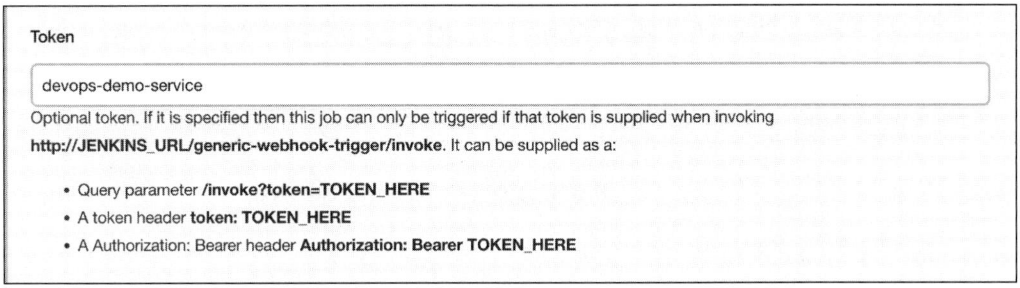

图 5-18　Token 配置

开启 Debug 调试模式，图 5-19 中的两个选项有助于排查集成问题。例如，集成时因为变量、参数导致的问题可以在 Jenkins Pipeline 构建日志中显示出来。

图 5-19　开启调试日志输出

3．触发测试

当我们配置好触发器后，可以使用 curl 命令进行测试触发，命令如下。

```
curl http://192.168.1.200:8080/generic-webhook-trigger/invoke?token=devops-demo-service
```

触发成功会有响应结果，代码如下。

```
{
    "jobs":{
        "devops-demo-service":{
            "regexpFilterExpression":"",
            "triggered":true,
            "resolvedVariables":{
                "WebHookData":""
            },
            "regexpFilterText":"",
            "id":56,
            "url":"queue/item/56/"
        }
    },
    "message":"Triggered jobs."
}
```

下面查看 Jenkins Pipeline 构建记录，如果触发成功，会显示出构建历史记录，如图 5-20 所示。

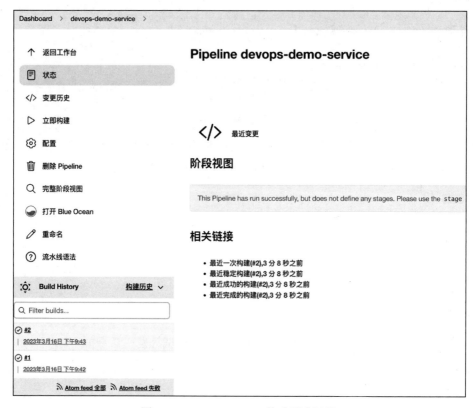

图 5-20　Jenkins Pipeline 构建历史记录

至此，我们完成了 GitLab 与 Jenkins 集成实践中 Jenkins 部分的配置。

5.3.2　GitLab 配置

GitLab 内置 Webhook 特性，可以在代码库开启并配置对应的事件以触发 Webhook。

1．项目开启 Webhook

进入代码库，导航到 Settings，单击左侧的 Webhooks 进入配置页面。在 URL 文本框中填写 Jenkins 的触发地址，在 Trigger 选项组中选择 Push events 触发事件，如图 5-21 所示。

可以看到当前代码库具有 Push events 时就会触发 Jenkins。代码提交、创建标签都属于 Push events。确认参数无误后，单击 Add webhook 按钮完成添加，如图 5-22 所示。

第 5 章　项目代码管理

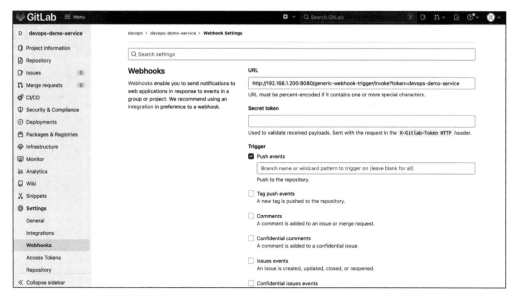

图 5-21　填写 Webhooks 参数

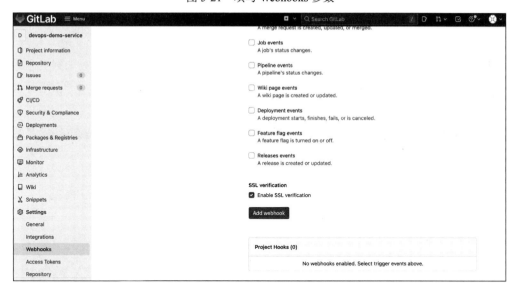

图 5-22　添加 webhook

如果提示"Url is blocked: Requests to the local network are not allowed"错误信息，需要开启 GitLab 的 local network 配置。导航到 Admin 管理页面，选择 Settings，单击 Network 进入配置页面。选中"Allow requests to the local network from web hooks and services"复选框。单击 Save changes 按钮保存配置，如图 5-23 所示。

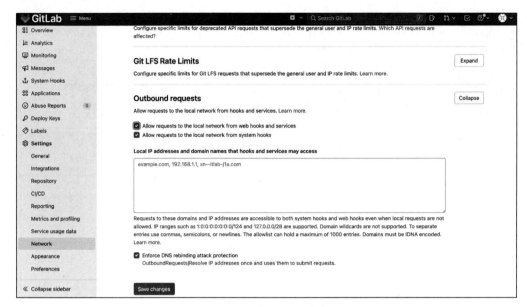

图 5-23　开启 local network 配置

2．测试 Webhook

进入代码库，导航到 Settings，单击 Webhooks 进入配置页面。将鼠标滚动到页面最下方可以看到已添加的 Webhook。在 Test 下拉列表框中选择 Push events 选项可以进行模拟触发，如图 5-24 所示。

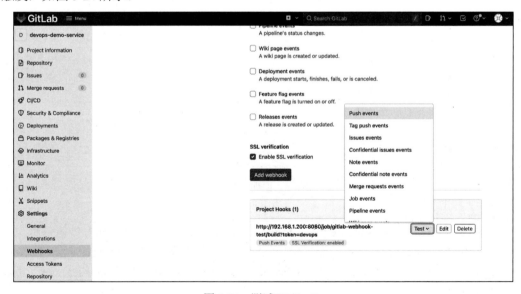

图 5-24　测试 Webhook

当出现"Hook executed successfully: HTTP 200"提示时表示执行成功。如果出现异常问题，可以单击 Edit 按钮进入 Webhook 排查，如图 5-25 所示。

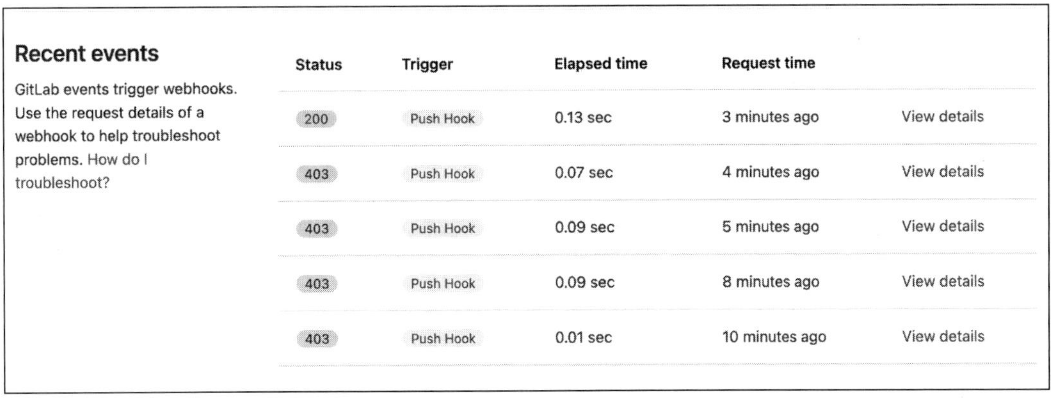

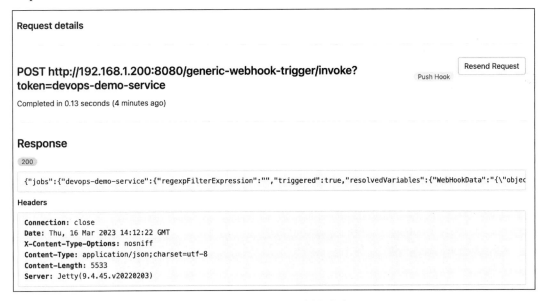

图 5-25　Webhook 触发记录

单击 View details 可以获取 GitLab Webhook 传递的数据类型和结构。假设因为环境网络等问题导致触发 Jenkins 失败，则不必重新提交代码来触发 Push event，可以单击 Resend Request 按钮重新进行触发，如图 5-26 所示。

图 5-26　Webhook 编辑页面

Request 部署的是 GitLab Webhook 传递给 Jenkins 的数据，如图 5-27 所示。

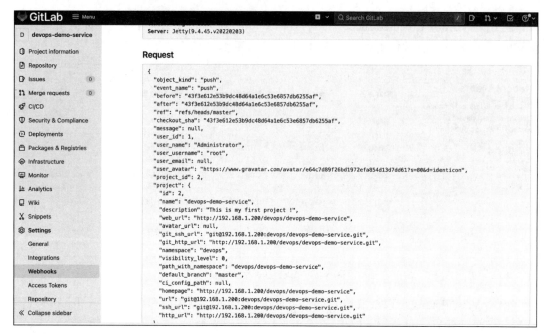

图 5-27　GitLab Webhook 数据

5.3.3　编写 Pipeline

1. 解析 Webhook 数据

编写 Jenkins Pipeline 读取 GitLab 参数并解析出分支名称和仓库地址。代码如下：

```
//GitLab 传递的数据{}
println("${WebHookData}")

//数据格式化
webHookData = readJSON text: "${WebHookData}"

//提取仓库信息
env.srcUrl = webHookData["project"]["git_http_url"]   //项目地址
env.branchName = webHookData["ref"] - "refs/heads/"   //分支
env.commitId = webHookData["checkout_sha"]            //提交 ID
env.commitUser = webHookData["user_username"]         //提交人
env.userEmail = webHookData["user_email"]             //邮箱

currentBuild.description = "Trigger by Gitlab \n branch: ${env.branchName}
//增加构建描述
currentBuild.displayName = "${env.commitId}"          //将构建 ID 变为提交 ID
```

将上述代码保存到 Jenkins Pipeline 中,然后在 GitLab 项目中提交代码,测试是否可以正常触发,如图 5-28 所示。

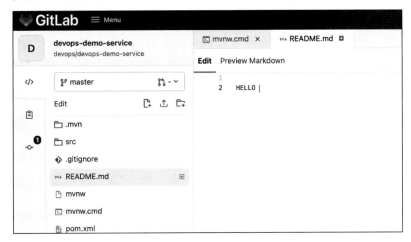

图 5-28　模拟提交代码

提交代码后,查看 Jenkins 构建结果。当出现了新的构建,说明配置成功,如图 5-29 所示。

图 5-29　Jenkins Pipeline 构建结果

2．下载代码

前面我们调试好了 Jenkins 解析 GitLab 传递的参数，接下来在 Pipeline 中添加下载代码。一般情况下，Jenkins Pipeline 中的第一个阶段就是下载代码。首先，在 Jenkins 系统中添加一个凭据，用于存储下载代码库的账户信息，如图 5-30 所示。

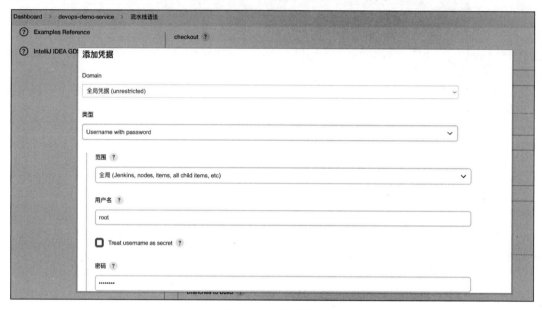

图 5-30　添加凭据

接下来是提交流水线中的核心逻辑——下载代码。打开片段生成器，生成下载功能的代码，如下所示。

```
//GitLab 传递的数据{}
println("${WebHookData}")

//数据格式化
webHookData = readJSON text: "${WebHookData}"

//提取仓库信息
env.srcUrl = webHookData["project"]["git_http_url"]       //项目地址
env.branchName = webHookData["ref"] - "refs/heads/"       //分支
env.commitId = webHookData["checkout_sha"]                //提交 ID
env.commitUser = webHookData["user_username"]             //提交人
env.userEmail = webHookData["user_email"]                 //邮箱

currentBuild.description = "Trigger by Gitlab \n branch: ${env.branchName}
```

```
//增加构建描述
currentBuild.displayName = "${env.commitId}"    //将构建 ID 变为提交 ID

pipeline {
    agent any
    stages {
        stage("CheckOut"){
            steps{
                script {
                    checkout([$class: 'GitSCM', branches: [[name: env.branchName ]],
                        extensions: [],
                        userRemoteConfigs: [[credentialsId: 'gitlab-admin-user', url: env.srcUrl ]]])
                    sh "ls -l"
                }
            }
        }
    }
}
```

在项目分支中提交代码,然后导航到 Jenkins 项目页面,可以看到正在运行的构建。如果配置成功,流水线会出现一个新的阶段——CheckOut,如图 5-31 所示。

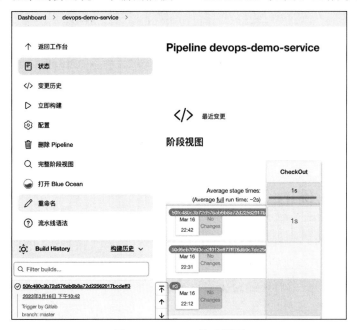

图 5-31　Jenkins 构建页面

单击"构建历史",查看 Jenkins 构建日志,可以发现下载代码的步骤已经生效。通过 ls 命令验证,发现已经成功地将代码下载到构建节点中,Jenkins 构建日志输出如图 5-32 所示。

```
[Pipeline] checkout
The recommended git tool is: NONE
using credential gitlab-admin-user
Cloning the remote Git repository
Cloning repository http://192.168.1.200/devops/devops-demo-service.git
 > git init /var/lib/jenkins/workspace/devops-demo-service # timeout=10
Fetching upstream changes from http://192.168.1.200/devops/devops-demo-service.git
 > git --version # timeout=10
 > git --version # 'git version 2.31.1'
using GIT_ASKPASS to set credentials
 > git fetch --tags --force --progress -- http://192.168.1.200/devops/devops-demo-service.git
+refs/heads/*:refs/remotes/origin/* # timeout=10
 > git config remote.origin.url http://192.168.1.200/devops/devops-demo-service.git # timeout=10
 > git config --add remote.origin.fetch +refs/heads/*:refs/remotes/origin/* # timeout=10
Avoid second fetch
 > git rev-parse origin/master^{commit} # timeout=10
Checking out Revision 50fc480c3b72d576ab6b8a72d22562017bcdeff3 (origin/master)
 > git config core.sparsecheckout # timeout=10
 > git checkout -f 50fc480c3b72d576ab6b8a72d22562017bcdeff3 # timeout=10
Commit message: "Update README.md"
First time build. Skipping changelog.
[Pipeline] sh
+ ls -l
total 28
-rwxr-xr-x 1 jenkins jenkins 10284 Mar 16 22:42 mvnw
-rw-r--r-- 1 jenkins jenkins  6734 Mar 16 22:42 mvnw.cmd
-rw-r--r-- 1 jenkins jenkins  1226 Mar 16 22:42 pom.xml
-rw-r--r-- 1 jenkins jenkins    14 Mar 16 22:42 README.md
drwxr-xr-x 4 jenkins jenkins    30 Mar 16 22:42 src
[Pipeline] }
```

图 5-32 Jenkins 构建日志输出

5.3.4 Pipeline 优化

可以对 Jenkins 的 Generic Webhook 进行优化,以避免出现不必要的构建。具体做法是在 Jenkins 中设置触发构建的条件,例如只有当指定分支更新时才触发构建,或者只有在特定标签添加时才构建。这样可以避免构建不必要的版本,同时也能够减轻服务器的负担,提高系统的稳定性和可靠性。

如果要过滤新建分支和标签的触发构建,可以在 Generic Webhook 中添加 3 个变量,分别是 object_kind、before、after，以获取当前的提交信息，如图 5-33 所示。

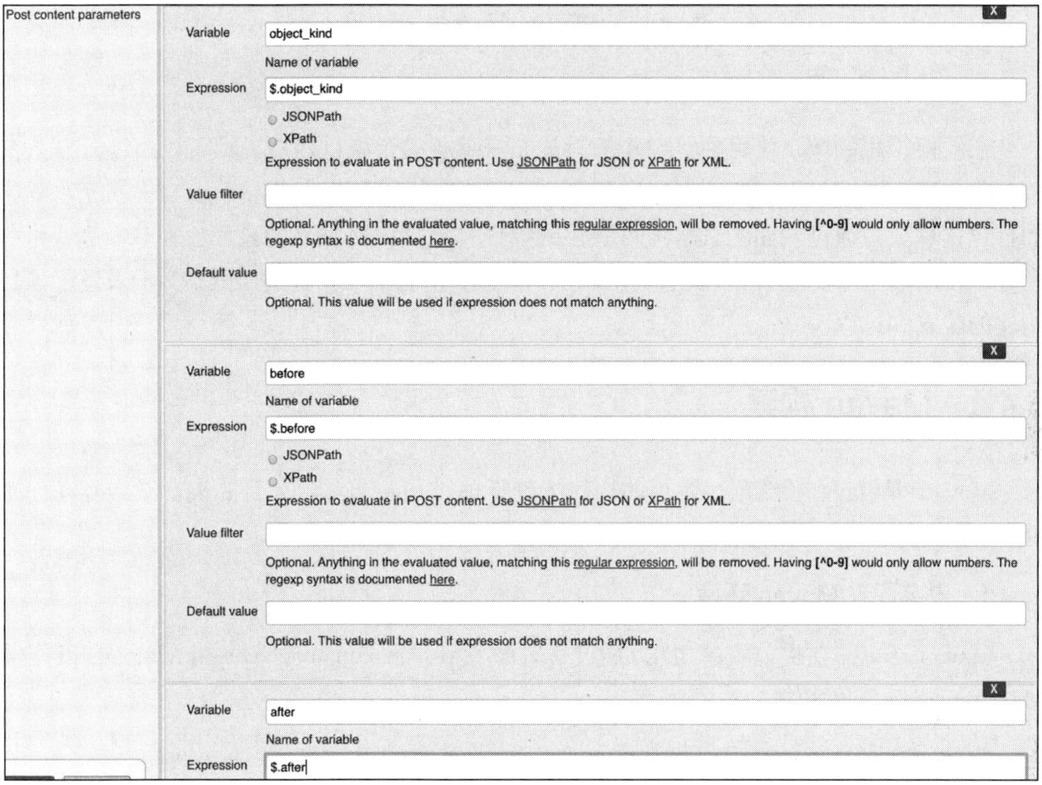

图 5-33　Generic Webhook 配置

通过正则表达式配置触发条件，在 Expression 文本框中填写 ^push\s(?!0{40}).{40}\s(?!0{40}).{40}$，在 Text 文本框中填写$object_kind $before $after。

该表达式匹配 push 请求只有在 after 和 before 的值都不是 40 个 0 时才会触发构建，当值为 40 个 0 时，删除分支或者新建分支操作，过滤条件配置如图 5-34 所示。

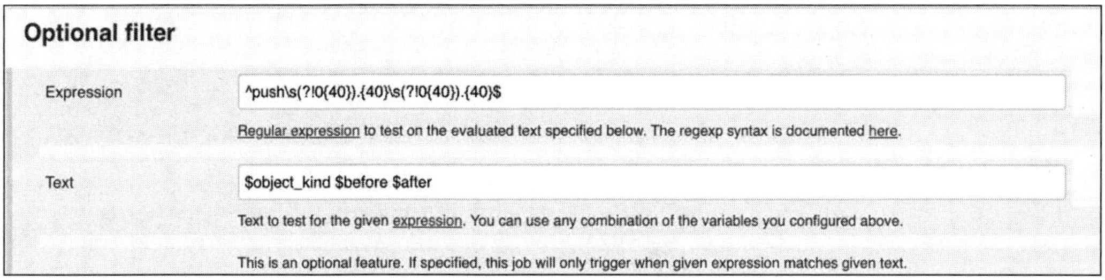

图 5-34　过滤条件配置

5.4 项目构建工具

在软件工程领域，使用项目构建工具是非常重要的，项目构建工具可以帮助开发人员自动化构建、测试、打包和部署应用程序，从而提高开发效率和质量。通常说的流水线中的打包过程就是通过构建工具将项目源代码编译打包后生成制品的过程。本节，笔者将分别介绍使用后端常用项目构建工具 Maven、Gradle 和前端项目构建工具 NPM 来进行项目构建的配置和过程。

5.4.1 Maven 构建

Maven 是 Java 后端项目常用的构建依赖管理工具。通常，项目的根目录会存在一个 pom.xml 文件，该文件用于定义项目的依赖包信息和构建配置。

1. 安装配置 Maven 环境

Maven 环境需要提前安装 JDK。通过官方源 https://maven.apache.org/download.cgi 下载 Maven 安装包。命令如下。

```
#下载安装包
wget https://mirrors.bfsu.edu.cn/apache/maven/maven-3/3.8.6/binaries/apache-maven-3.8.5-bin.tar.gz
#解压安装包
tar zxf apache-maven-3.8.5-bin.tar.gz -C /usr/local/
cd /usr/local/apache-maven-3.8.6/

#配置环境变量
vi /etc/profile
export M2_HOME=/usr/local/apache-maven-3.8.6
export PATH=$M2_HOME/bin:$PATH
source /etc/profile

#验证
#mvn -v
Apache Maven 3.8.6 (84538c9988a25aec085021c365c560670ad80f63)
Maven home: /data/devops5/tools/apache-maven-3.8.6
Java version: 1.8.0_201, vendor: Oracle Corporation, runtime: /usr/local/jdk1.8.0_201/jre
Default locale: zh_CN, platform encoding: UTF-8
```

```
OS name: "linux", version: "4.18.0-373.el8.x86_64", arch: "amd64", family:
"unix"
```

2. 创建 Maven 项目

访问 https://start.spring.io/ 生成一个 Maven 类型的项目。Project 的类型选择 Maven，Spring Boot 版本可以自定义，这里选择 2.7.10。Project Metadata 用于记录元数据信息，保持默认选项即可。单击 GENERATE 按钮创建项目，如图 5-35 所示。

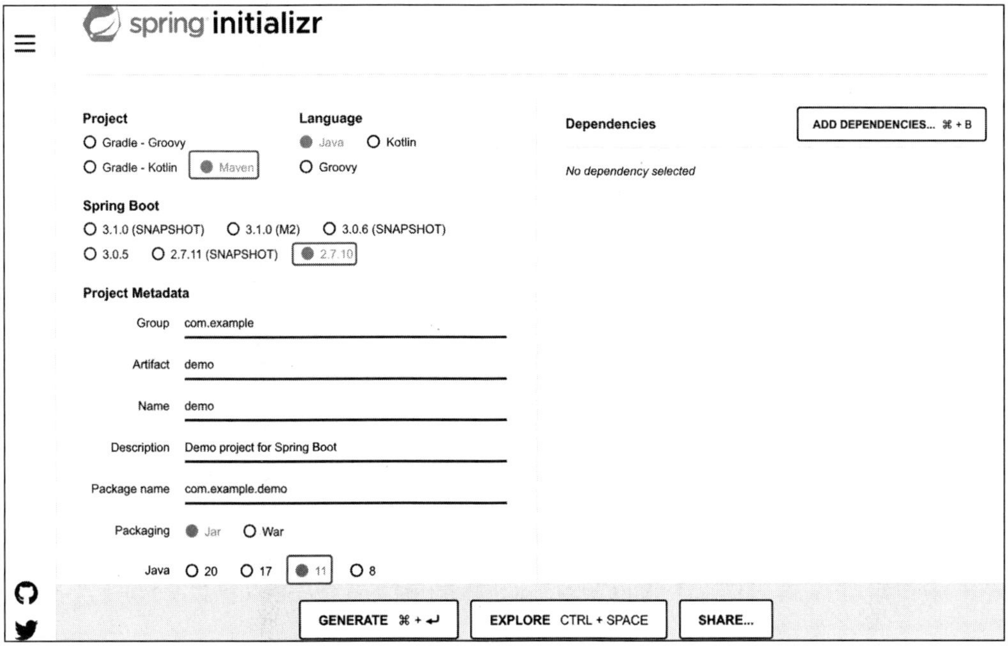

图 5-35　创建 Maven 项目

3. 构建 Maven 项目

Maven 具有一系列的命令选项和参数，可以自动化构建应用程序，包括编译代码、打包应用程序和安装必要的依赖项。下面笔者整理了一些常用的命令参数。

```
#清理构建目录和缓存
mvn clean
#项目打包
mvn clean package
#项目打包并部署本地
mvn clean install
#进行单元测试
mvn clean test
```

```
#指定配置文件
mvn clean package -f ../pom.xml
#运行打包并跳过单元测试
mvn clean package -DskipTests
mvn claan package -Dmaven.test.skip=true
#将包发布到远程制品库
mvn deploy
```

以上是常用的 Maven 构建命令，企业中根据实际情况参数会略有不同。

4. Jenkins 集成

Maven 可以运行编译命令进行项目构建。那么在企业中实施 Jenkins Pipeline 时如何集成呢？实际上，在 Jenkins 流水线中集成 Maven 并不复杂，类似于执行一条 Shell 命令。流水线代码如下。

```
//Jenkins Pipeline 中的 Build 阶段
stage("Build"){
steps{
        script{
            //sh 执行 Shell 命令
            sh "mvn clean package"
        }
    }
}
```

如果命令执行失败，读者可以尝试手动运行命令在环境中测试。一般情况下，如果 Shell 命令执行失败，可以暂时排除 Jenkins 自身的问题而在构建节点环境和项目源代码方向排错。

5.4.2 Gradle 构建

Gradle 同样是 Java 后端项目中常用的构建依赖管理工具，与 Maven 相比，编译和依赖解析更快。

1. 安装配置 Gradle 环境

Gradle 环境同样需要提前安装 JDK。通过官方源 https://gradle.org/releases/ 下载 Gradle 安装包。安装命令如下。

```
#解压安装包
unzip gradle-7.5.1-bin.zip -d /usr/local/

#配置环境变量
```

```
vi /etc/profile
export GRADLE_HOME=/usr/local/gradle-7.5.1/
export PATH=$GRADLE_HOME/bin:$PATH
source /etc/profile

#验证
gradle -v
```

2. 创建 Gradle 项目

访问 https://start.spring.io/ 生成一个 Gradle 类型的项目。Project 类型选择 Gradle，Spring Boot 版本可以自定义，这里选择 2.7.10。Project Metadata 用于记录元数据信息，保持默认选项即可。单击 GENERATE 创建项目，如图 5-36 所示。

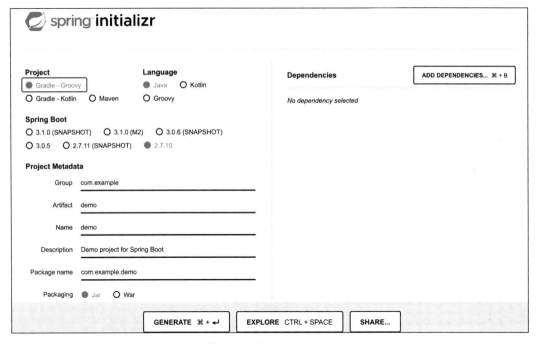

图 5-36　创建 Gradle 项目

3. 构建 Gradle 项目

Gradle 工具同样具有一系列的命令选项和参数，可以自动化构建应用程序，包括编译代码、打包应用程序和安装必要的依赖项。常用的命令参数如下。

```
#清理构建目录和缓存
gradle clean
#构建项目
```

```
gradle build
#构建项目并跳过测试
gradle build -x test
```

以上是常用的 Gradle 构建命令,企业中根据实际情况参数可以略有不同。

4. Jenkins 集成

在 Jenkins 流水线中集成 Gradle 并不复杂,类似于执行一条 Shell 命令。流水线代码如下。

```
//Jenkins Pipeline 中的 Build 阶段
stage("Build"){
steps{
        script{
            //sh 执行 shell 命令
            sh "gradle build"
        }
    }
}
```

如果命令执行失败,读者可以尝试手动运行命令在环境中测试。如果 Shell 命令执行失败,读者可以在构建节点环境和项目源代码方向排错。

5.4.3 NPM 构建

Node.js Package Manager(NPM)是一种用于管理 Node.js 应用程序依赖关系的脚本工具。它是使用最广泛的 Node.js 包管理器之一,可以帮助开发人员更方便地安装、更新、部署和管理 Node.js 应用程序。

1. 安装配置 NPM 环境

NPM 是命令行工具,需要提前安装 Node.js 环境。首先需要通过官方源 https://nodejs.org/en/download/ 下载安装包。安装命令如下。

```
#下载安装包
wget https://nodejs.org/dist/v14.16.1/node-v14.16.1-linux-x64.tar.xz

#解压安装包
tar xf node-v14.16.1-linux-x64.tar.xz -C /usr/local/

#修改环境变量
#vi /etc/profile
export NODE_HOME=/usr/local/node-v14.16.1-linux-x64
export PATH=$NODE_HOME/bin:$PATH
```

```
#source /etc/profile

#验证
#node -v
v14.16.1
#npm -v
6.14.12
```

2．创建 NPM 项目

Vue.js 是一种流行的前端 JavaScript 框架，在开发 Vue.js 项目时，使用 NPM 可以帮助开发人员更方便地安装、更新、部署和管理 Vue.js 应用程序。创建 Vue.js 项目的命令如下。

```
#安装 vue-cli 工具
#npm install -g vue-cli
+ vue@2.6.12
added 1 package from 1 contributor in 3.342s
#初始化 Vue 项目
#vue-init  webpack vuedemo

? Project name vuedemo
? Project description A Vue.js project
? Author adminuser <xxxxx@xxxx.com>
? Vue build standalone
? Install vue-router? No
? Use ESLint to lint your code? No
? Set up unit tests No
? Setup e2e tests with Nightwatch? No
? Should we run `npm install` for you after the project has been created? (recommended) npm
   vue-cli · Generated "vuedemo".
```

3．构建 NPM 项目

NPM 工具同样具有一系列的命令选项和参数，可以自动化构建应用程序，包括编译代码、打包应用程序和安装必要的依赖项。笔者整理了一些常用的命令参数。

```
#安装模块
npm install <moduleName> -g
#查看已安装包
npm list
#配置使用淘宝源
npm config set registry https://registry.npm.taobao.org
#配置模块缓存路径
npm config set cache  "/opt/npmcache/"
```

```
#构建
npm run build
```

以上是常用的 NPM 构建命令,企业中根据实际情况参数可以略有不同。

4. Jenkins 集成

Jenkins 流水线中集成 NPM 并不复杂,执行一条 Shell 命令。流水线代码如下。

```
//Jenkins Pipeline 中的 build 阶段
stage("Build"){
steps{
        script{
            //sh 执行 shell 命令
            sh "npm install && npm run build"
        }
    }
}
```

如果命令执行失败,读者可以尝试手动运行命令在环境中测试。如果 Shell 命令执行失败,读者可以在构建节点环境和项目源代码方向排错。

5.5 本章小结

本章主要讲解了 GitLab 版本控制系统入门和配置提交流水线。为读者展示了通过使用 Jenkins Generic Webhook 插件和 GitLab 集成实现开发人员提交代码后,Pipeline 自动触发并获取项目代码库和分支信息。至此,读者已经掌握了 Jenkins 和 GitLab 版本控制系统集成的实践能力。

第 6 章 代码质量平台实战

SonarQube 是一个开源的、基于 Java 的代码质量分析工具，旨在为软件开发者和质量保证人员提供对代码和项目的全面分析和监控。它提供了对代码质量、代码重复、依赖关系、代码复杂度等方面的分析和监控，可以帮助开发者更好地管理代码质量和风险。本章笔者将介绍 SonarQube 平台在企业中的实践，实践环境采用最新的是长期支持版本 9.9。

本章内容速览。

- ☑ SonarQube 系统入门
- ☑ SonarQube 代码扫描
- ☑ SonarQube 系统集成
- ☑ 本章小结

6.1　SonarQube 系统入门

6.1.1　SonarQube 概述

SonarQube 是一款代码审查工具，可以帮助我们交付整洁干净的代码，支持 30 多种语言和框架并逐渐成为软件质量分析行业的标准。SonarQube 提供命令行和插件的方式，支持与 DevOps 系统集成，如 Jenkins、GitLab 等平台，如图 6-1 所示。

SonarQube 是一个服务器/客户端架构，SonarQube 组件组成如图 6-2 所示，它由 3 个主要组件组成：SonarQube Server、Database Server 和 Scanner。SonarQube Server 是一个开源的 Java 应用程序，它管理并运行实例或集群，Scanner 是通过 Web 界面或命令行界面与服务器进行交互的应用程序。Scanner 提供了对 SonarQube 数据库的访问，并允许用户进行数据分析、报告生成和配置等操作。

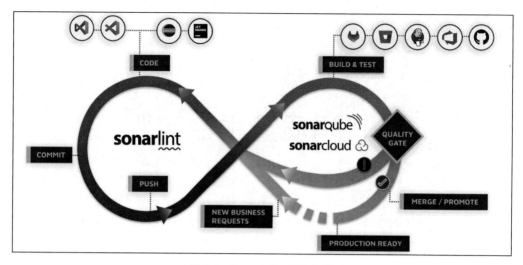

图 6-1　SonarQube 产品介绍

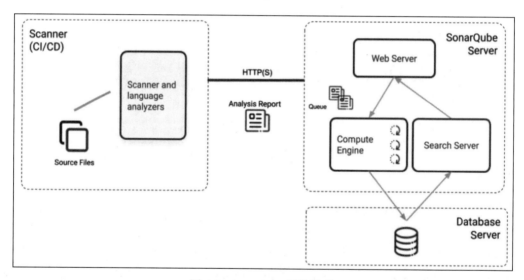

图 6-2　SonarQube 组件组成

（1）SonarQube Server 有 3 个主要进程。① Web Server：提供 Web 页面的服务，以供开发人员查看软件质量报告和配置实例。② Search Server：提供 Web 页面中的搜索服务，基于 Elasticsearch。③ Compute Engine：计算引擎负责处理代码分析报告并将其保存在数据库中。

（2）Database Server 用于存储平台的数据。① 源代码文件：包括代码文件的路径、大小、修改次数、版本控制信息等；② 代码分析结果：代码静态分析，检测代码中的潜在问题和缺陷，并将这些数据存储在数据库中，包括代码质量指标、代码规则、问题数量和严重程度等。

（3）Scanner 是 SonarQube 的内置扫描工具，用于扫描源代码和依赖项，并将生成的代码中存在问题和缺陷的报告存入数据库。

6.1.2　SonarQube 安装

本节我们将使用 Docker 安装 SonarQube，该方式便于我们快速部署和测试。如果在生产环境部署，建议通过 RPM 包或者基于 Kubernetes 部署的方式使数据库与容器分离。注意：新版本的 SonarQube 所依赖的 JDK 环境版本应在 17 以上，数据库可以采用 PostgreSQL，如图 6-3 所示。

为避免容器删除后数据丢失，我们在本地创建需要持久化的目录。具体而

Java	Server	Scanners
Oracle JRE	✓ 17	✓ 17
	✗ 11	✓ 11
OpenJDK	✓ 17	✓ 17
	✗ 11	✓ 11

图 6-3　SonarQube 支持的 Java 版本图

言，需要创建 conf、extensions、logs、data 目录，将其挂载到 Docker 容器中，以完成 SonarQube 服务的启动及数据的持久化，命令如下。

```
mkdir -p /data/devops/sonarqube/{sonarqube_conf,sonarqube_extensions}
mkdir -p /data/devops/sonarqube/{sonarqube_logs,sonarqube_data}
```

在启动容器时，通过 Docker 命令行工具中的-v 参数，将本地目录与容器内的数据目录挂载起来，这样可以将 SonarQube 的数据持久化到本地磁盘，以防止容器删除后数据丢失。命令如下。

```
docker run -itd --name sonarqube \
   -p 9000:9000 \
   -v /data/devops/sonarqube/sonarqube_conf:/opt/sonarqube/conf \
   -v /data/devops/sonarqube/sonarqube_extensions:/opt/sonarqube/extensions \
   -v /data/devops/sonarqube/sonarqube_logs:/opt/sonarqube/logs \
   -v /data/devops/sonarqube/sonarqube_data:/opt/sonarqube/data \
   sonarqube:lts-community
```

访问 http://主机 IP:9000 进入 SonarQube 页面。自 8.x.x 版本开始，SonarQube 加强了安全设置，需要验证后才能进入系统。默认账号为 admin，密码为 admin，如图 6-4 所示。

第一次登入时需要更新密码，更新密码页面如图 6-5 所示。

图 6-4　SonarQube 登入页面　　　　　图 6-5　更新密码页面

　　SonarQube 平台可以支持与 DevOps 平台集成。创建 SonarQube 项目的方式有很多种，例如可以配置 From GitLab 导入已存在的项目，还可以使用 Manually 的方式手动创建项目。SonarQube 创建项目页面如图 6-6 所示。

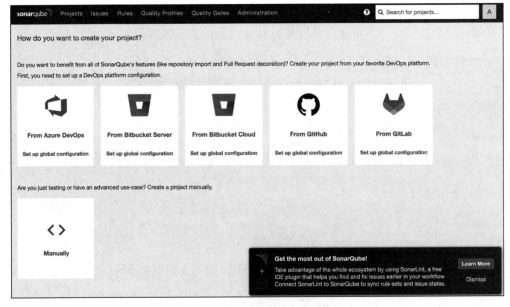

图 6-6　SonarQube 创建项目页面

6.1.3 插件管理

SonarQube 可以通过插件来扩展功能，但 SonarQube 官方并不提供插件。因此，安装风险由用户自行承担。SonarQube 官方对安装和使用的插件不承担任何责任。用户可以在单击 I understand the risk 按钮确认风险后直接从下面的列表中安装插件，如图 6-7 所示。

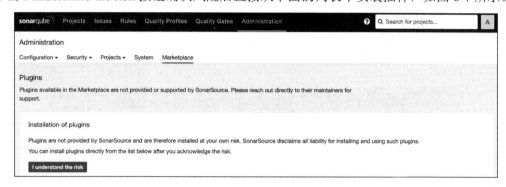

图 6-7　确认插件安装风险

搜索要安装的插件，这里我们可以选择安装 Chinese Pack（中文插件）来将页面从英文模式变成中文模式。单击 Install 按钮进行安装即可。需要注意的是，安装插件时需要容器的 SonarQube 服务，所以最好在合适的时间安装插件，避免造成服务中断，如图 6-8 所示。

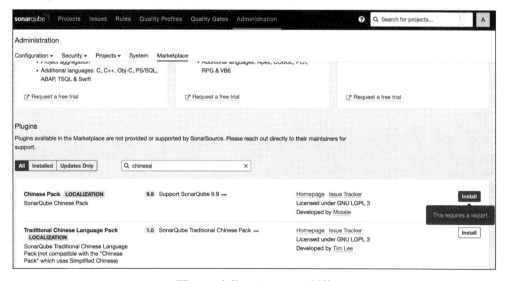

图 6-8　安装 Chinese Pack 插件

安装插件的第一步是将插件下载到 SonarQube 的插件目录中。进入容器可以找到插件的下载目录。查找目录的命令如下。

```
[root@devops-nuc-service ~]#docker exec -it sonarqube bash
sonarqube@4a72a77480b4:/opt/sonarqube$ ls
sonarqube@4a72a77480b4:/opt/sonarqube$ cd extensions/
sonarqube@4a72a77480b4:/opt/sonarqube/extensions$ ls
downloads
sonarqube@4a72a77480b4:/opt/sonarqube/extensions$ ls downloads/
sonar-l10n-zh-plugin-9.9.jar
```

插件下载成功后服务器页面会提示需要重启服务。单击 Restart Server 按钮完成服务重启，如图 6-9 所示。

图 6-9 提示重启服务界面

插件安装成功后，经过重启服务页面从英文模式变成了中文模式，如图 6-10 所示。

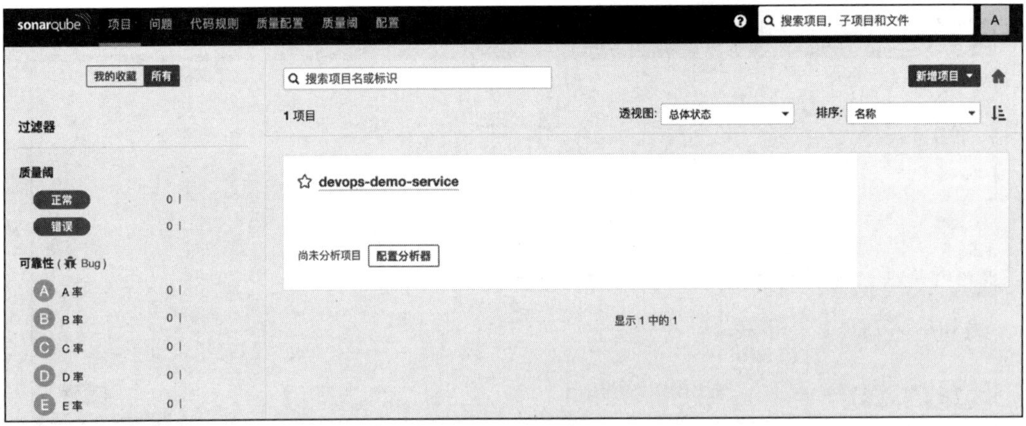

图 6-10 安装中文插件后的首页

在企业中，网络策略有严格的访问限制，一般不能用服务器直接连接公网。所以在内网的情况下我们并不能访问公网。那么该如何安装插件呢？我们可以选择在公网下载插件，然后手动将其上传到 SonarQube 服务的插件目录<SONARQUBE_HOME>/extensions/plugins，再重启 SonarQube 服务。

6.2　SonarQube 代码扫描

SonarQube 最多可以分析 29 种不同的语言，具体取决于软件的版本，分析的内容会因语言而异。在所有语言上，都对源代码进行静态分析。对于某些语言，静态分析应该在编译后的代码上完成，如 Java 中的.class 文件和 C#中的.dll 文件等。

6.2.1　SonarQube 质量配置

SonarQube 中每种编程语言都会有一些内置的代码规则。代码规则是 SonarQube 对源代码进行分析的基础，它们定义了如何扫描代码并识别潜在的问题。SonarQube 进行代码质量检查也是通过这些规则来判断的，代码规则页面如图 6-11 所示。

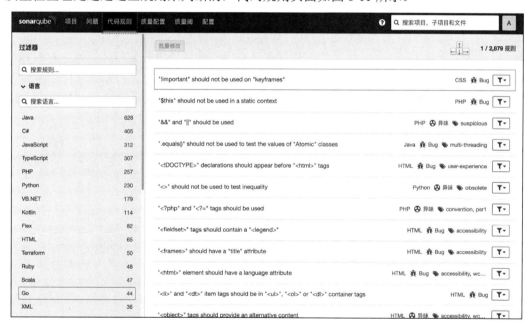

图 6-11　代码规则页面

质量配置则定义了如何对这些潜在的问题进行评估，以及如何根据评估结果调整扫描规则。在企业中，每个公司或者组织对质量配置要求定义不一致，可以单击"创建"按钮来创建质量配置以自定义质量标准，如图 6-12 所示。

图 6-12　质量配置页面

6.2.2　SonarQube 质量阈

质量阈指的是质量门禁，通常作为提交流水线中的质量关卡及项目的质量。质量阈也可以根据公司和组织的侧重点来进行合理设置。在设置质量阈值时，可以定义一个质量度量条件，如代码覆盖率，并将其设置为一个阈值。如果代码中的覆盖率小于该阈值，则质量不合格。这样，就可以对代码质量进行管控，如图 6-13 所示。

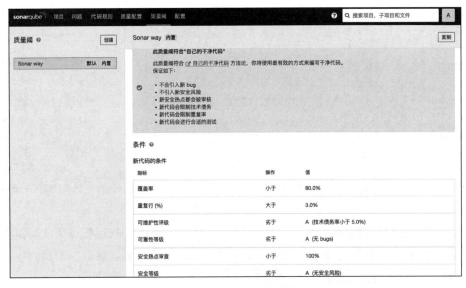

图 6-13　SonarQube 质量阈

6.2.3 Sonar Scanner 配置

SonarQube 平台使用 Sonar Scanner 进行代码扫描。Sonar Scanner 为不同项目构建工具提供了插件，如 Maven、Gradle、.NET、Jenkins 等，还提供了命令行工具。一般代码扫描阶段需要和提交流水线集成，所以这里我们重点讲解关于通用的命令行工具和 Jenkins 插件的使用方式。

1. 安装 Scanner

Scanner 的安装需要下载安装包，并将包路径加入系统环境变量。命令如下。

```
##下载包
wget https://binaries.sonarsource.com/Distribution/sonar-scanner-cli/sonar-scanner-cli-4.8.0.2856-linux.zip

##解压包
unzip sonar-scanner-cli-4.8.0.2856-linux.zip

##设置环境变量
vim /etc/profile
export SONAR_SCANNER_HOME=/root/books/sonar-scanner-4.8.0.2856-linux
export PATH=$SONAR_SCANNER_HOME/bin:$PATH
source /etc/profile

##测试生效
sonar-scanner -v
```

配置正确后，可以通过 sonar-scanner -v 命令查看当前的版本信息，如图 6-14 所示。

```
[root@devops-nuc-service books]# cd sonar-scanner-4.8.0.2856-linux/
[root@devops-nuc-service sonar-scanner-4.8.0.2856-linux]# pwd
/root/books/sonar-scanner-4.8.0.2856-linux
[root@devops-nuc-service sonar-scanner-4.8.0.2856-linux]# cd ..
[root@devops-nuc-service books]# sonar-scanner -v
INFO: Scanner configuration file: /root/books/sonar-scanner-4.8.0.2856-linux/conf/sonar-scanner.properties
INFO: Project root configuration file: NONE
INFO: SonarScanner 4.8.0.2856
INFO: Java 11.0.17 Eclipse Adoptium (64-bit)
INFO: Linux 4.18.0-373.el8.x86_64 amd64
```

图 6-14 查看当前版本信息

2. 扫描参数配置

Sonar Scanner 扫描项目时需要一些项目参数，如项目的关键字、项目名称、项目版本包、代码扫描目录、语言编码、SonarQube 服务端信息等，具体如下。

```
#定义唯一的关键字
sonar.projectKey=devops-hello-service

#定义项目名称
sonar.projectName=devops-hello-service

#定义项目的版本信息
sonar.projectVersion=1.0

#指定扫描代码的目录位置（多个目录需要逗号分隔）
sonar.sources=.

#执行项目编码
sonar.sourceEncoding=UTF-8

#指定 sonar Server
sonar.host.url=
sonar.login=
sonar.password=
```

参数的传递有两种方式，可以通过写入文件，也可以通过命令行直接传递。默认情况下会加载名称为 sonar-project.properties 的参数文件。使用方式如下。

```
#指定配置文件
sonar-scanner -Dproject.settings=myproject.properties

#命令行传参
sonar-scanner -Dsonar.projectKey=myproject -Dsonar.sources=src1
```

3. 扫描 Java 项目

我们需要在 https://start.spring.io/ 下生成一个 Java 项目代码到本地，然后使用 Sonar Scanner 进行扫描。测试项目是用 Java 语言编写的，采用的是 Spring Boot 框架，项目构建工具是 Apache Maven，如图 6-15 所示。

Java 项目有个特点，如果需要对 .class 文件进行扫描，就需要先编译构建项目，然后再进行代码扫描。因为只有编译后才能获取 .class 文件。下载项目的命令如下。

```
#上传项目到远程服务器
scp Downloads/demo.zip root@192.168.1.200:/root/books

#登入远程服务器编译构建
```

```
unzip demo.zip
cd demo /
mvn clean package
```

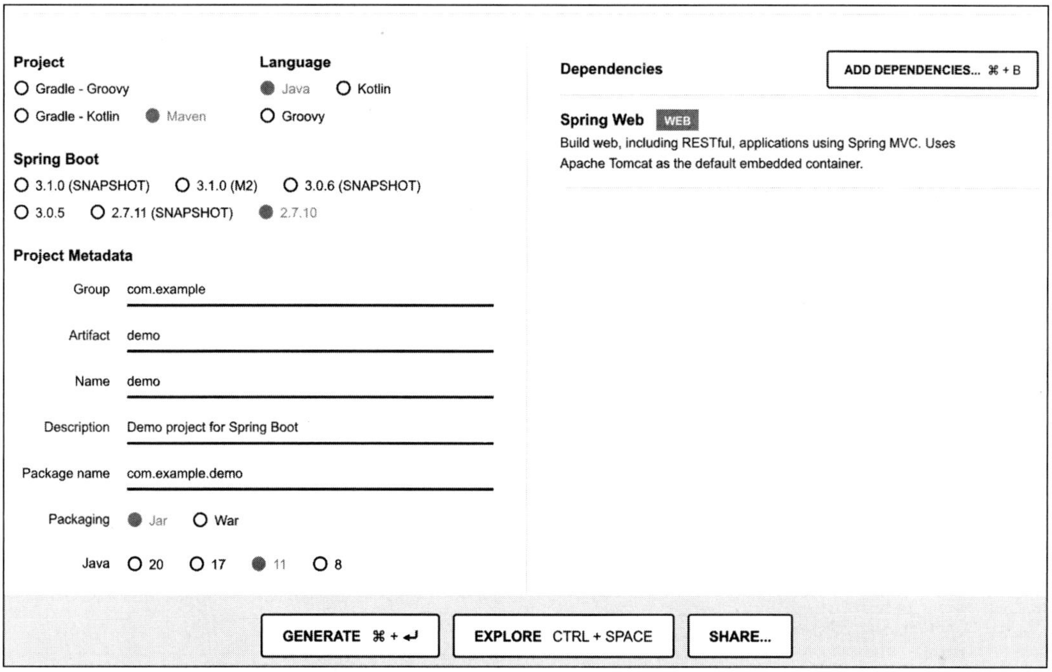

图 6-15　测试的 Java 项目

编译成功后会出现一个 target 目录，target/classes 目录用于存放 .class 文件。当显示"BUILD SUCCESS"时表示项目构建成功，如图 6-16 所示。

```
-lang3/3.7/commons-lang3-3.7.jar (500 kB at 372 kB/s)
[INFO] Replacing main artifact with repackaged archive
[INFO] ------------------------------------------------------------------------
[INFO] BUILD SUCCESS
[INFO] ------------------------------------------------------------------------
[INFO] Total time:  52.656 s
[INFO] Finished at: 2023-04-09T10:17:13+08:00
[INFO] ------------------------------------------------------------------------
```

图 6-16　项目构建成功

编写扫描参数文件 sonar-project.properties，内容如下。

```
#项目关键字
sonar.projectKey=devops-hello-service

#项目名称
sonar.projectName=devops-hello-service
```

```
#项目的版本信息
sonar.projectVersion=1.0

#指定扫描代码的目录位置
sonar.sources=src

#执行项目编码
sonar.sourceEncoding=UTF-8

#指定sonar Server
sonar.host.url=http://192.168.1.200:9000
#认证信息
sonar.login=admin
sonar.password=admin123

#Java class 目录
sonar.java.binaries=target/classes
sonar.java.test.binaries=target/test-classes
sonar.java.surefire.report=target/surefire-reports
```

运行 sonar-scanner 命令可以直接加载配置文件并运行扫描。扫描成功后，本地会出现下面的提示日志，如图 6-17 所示。

```
INFO: ANALYSIS SUCCESSFUL, you can find the results at: http://192.168.1.200:9000/dashboard?id=devops-hello-service
INFO: Note that you will be able to access the updated dashboard once the server has processed the submitted analysis report
INFO: More about the report processing at http://192.168.1.200:9000/api/ce/task?id=AYdj1IFd910gCQ59GOXC
INFO: Analysis total time: 7.813 s
INFO: ------------------------------------------------------------------------
INFO: EXECUTION SUCCESS
INFO: ------------------------------------------------------------------------
INFO: Total time: 9.000s
INFO: Final Memory: 20M/88M
INFO: ------------------------------------------------------------------------
```

图 6-17 扫描成功日志

此时，可以在 SonarQube 服务器中看到扫描项目的质量信息，如图 6-18 所示。

图 6-18 SonarQube 项目的质量信息

6.3　SonarQube 系统集成

在持续集成流水线中，代码质量检查阶段是很重要的部分。本节，笔者将根据常用的两种方式进行讲解。

6.3.1　准备工作

将 7.2.3 节中的演示项目上传到 GitLab 系统。在 GitLab 中新建项目，名称为 demo-sonar-service，然后将本地的源码上传到 GitLab 项目中。上传命令如下。

```
git init --initial-branch=main
git remote add origin http://192.168.1.200/devops/demo-sonar-service.git
git add .
git commit -m "Initial commit"
git push -u origin main
```

上传成功后，进入 GitLab 查看，项目首页图如图 6-19 所示。

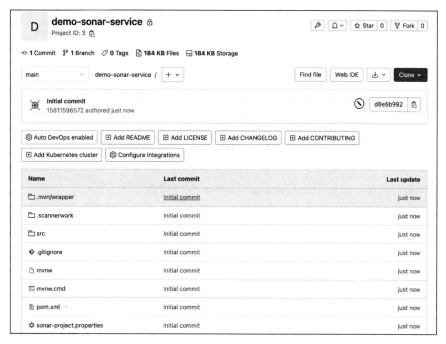

图 6-19　项目首页图

为了安全地把 sonar-project.properties 文件中的 sonar.login 和 sonar.password 参数删除，生产环境不应该把敏感数据放到代码中。删除后的内容如下。

```
#定义唯一的关键字
sonar.projectKey=devops-sonar-service

#定义项目名称
sonar.projectName=devops-sonar-service

#定义项目的版本信息
sonar.projectVersion=1.0

#指定扫描代码的目录位置（多个逗号分隔）
sonar.sources=src

#执行项目编码
sonar.sourceEncoding=UTF-8

#指定 sonar Server
sonar.host.url=http://192.168.1.200:9000

#认证信息
#sonar.login=xxx
#sonar.password=xxxx

#Java class 目录
sonar.java.binaries=target/classes
sonar.java.test.binaries=target/test-classes
sonar.java.surefire.report=target/surefire-reports
```

将 SonarQube 系统的认证用户信息存储到 Jenkins 凭据中，如图 6-20 所示。新建凭据之后，会出现凭据 ID，在后面的 Jenkins Pipeline 中会用到这个 ID。

6.3.2 命令行方式

1. 创建 Pipeline

在 Jenkins 中新建流水线项目，名称为 demo-sonar-service，如图 6-21 所示。

第 6 章 代码质量平台实战

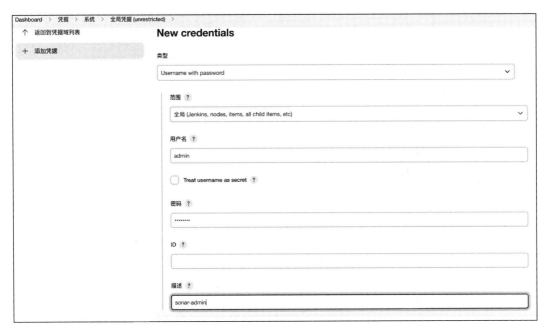

图 6-20 创建 Jenkins 凭据

图 6-21 创建 Pipeline

在项目设置中开启参数化构建并添加参数，如仓库地址、分支名称、构建命令等。Jenkins 项目参数页面如图 6-22 所示。

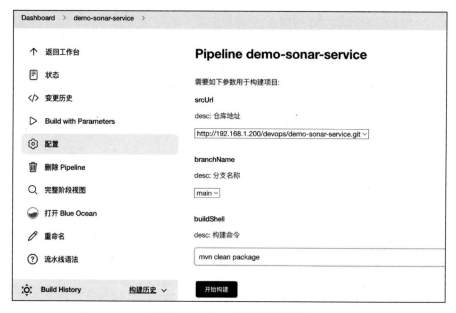

图 6-22　Jenkins 项目参数页面

2．编写 Jenkinsfile

编写 Jenkinsfile 分 3 个步骤：下载代码、构建代码、代码质量检查（见图 6-23）。

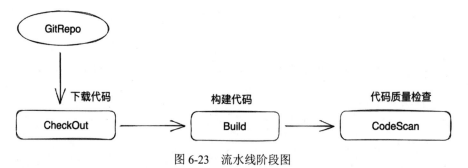

图 6-23　流水线阶段图

我们将图 6-23 的流水线阶段图转换成 Groovy 代码，具体如下。

```
//流水线
pipeline {
    //指定构建节点
    agent { label "build"}
    stages{
```

```
//下载代码
stage("CheckOut"){
    steps{
        script{
            println("CheckOut")
            checkout([$class: 'GitSCM',
                branches: [[name: "${params.branchName}"]],
                extensions: [],
                userRemoteConfigs:[[credentialsId:'gitlab-admin-user',
                url: "${params.srcUrl}"]]])
            sh "ls -l "  //验证

        }
    }
}
//构建代码
stage("Build"){
    steps{
        script{
            println("Build")
            sh "${params.buildShell}"
        }
    }
}
//代码扫描
stage("CodeScan"){
    steps{
        script{
            println("CodeScan")
            withCredentials([usernamePassword(credentialsId:
'702cae96-20f6-42dc-b8c2-d5a5f005e297',
                            passwordVariable: 'PASSWORD',
                            usernameVariable: 'USERNAME')]) {

            sh "sonar-scanner \
                -Dsonar.login=${USERNAME} \
                -Dsonar.password=${PASSWORD}"
            }
        }
    }
}
}
```

代码中使用 withCredentials 来读取 Jenkins 凭据并将用户名和密码读取到环境变量中，

以便于后续步骤使用。

3. 验证测试

单击"开始构建"按钮构建流水线，如图 6-24 所示。

图 6-24　构建流水线

如果构建失败，可以进一步通过日志进行排查。构建成功页面如图 6-25 所示。

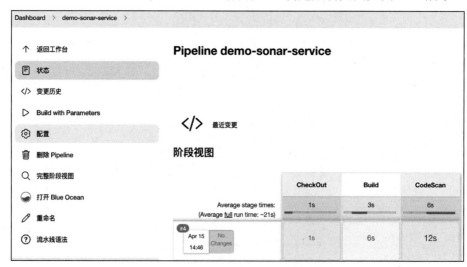

图 6-25　构建成功页面

登入 SonarQube 可以看到 demo-sonar-service 的项目质量结果，如图 6-26 所示。

图 6-26　项目质量结果

至此，我们完成了通过命令行的方式与 Jenkins 集成并运行代码扫描实践。

6.3.3　Jenkins 插件

SonarQube 官方提供了 Jenkins 集成插件，通过该插件可以实现 Jenkins 代码扫描。

1. 安装 SonarQube 插件

进入 Jenkins 插件管理，搜索 SonarQube。根据结果选择 SonarQube Scanner 插件并安装，安装 SonarQube 插件如图 6-27 所示。

图 6-27　安装 SonarQube 插件

2. 配置 SonarQube

导航到 Jenkins 系统配置页面，向下滚动到 SonarQube 配置部分，单击 Add SonarQube 按钮，添加服务器。填写 Name 和 Server URL，如图 6-28 所示。

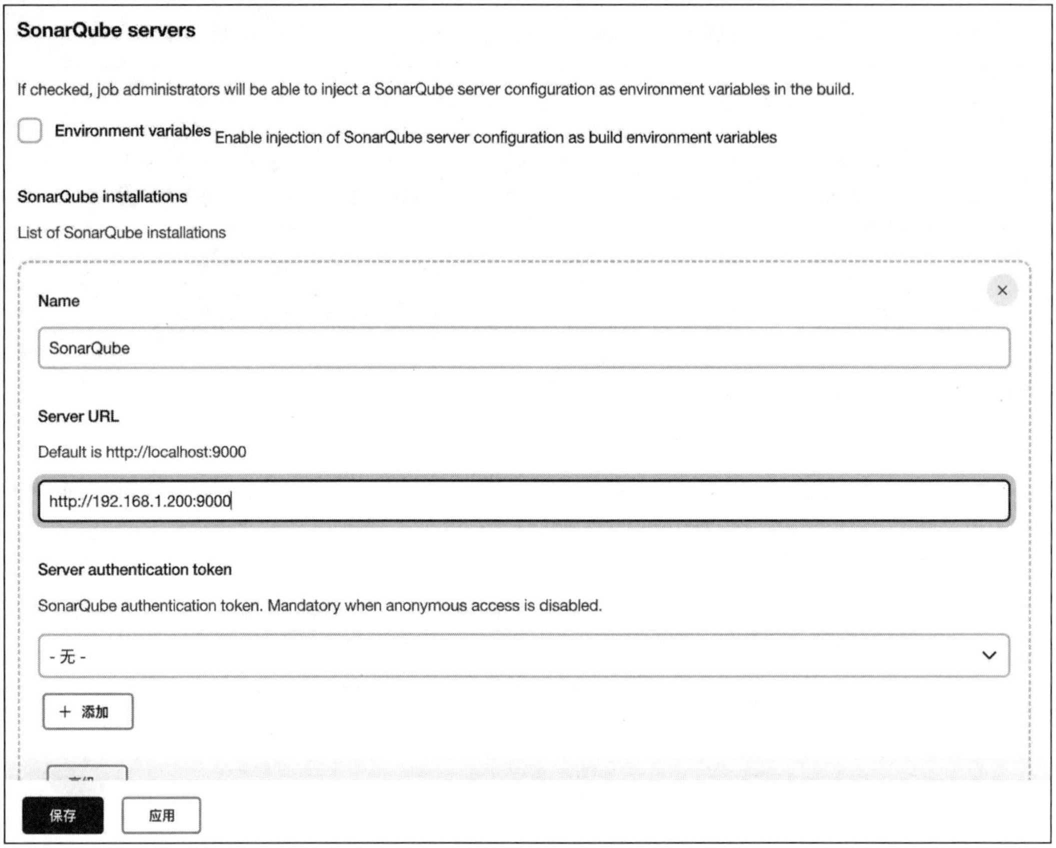

图 6-28　配置插件

3. 编写 Jenkinsfile

使用 SonarQube Scanner 插件在 Jenkins Pipeline 中扫描代码。代码如下。

```
//流水线
pipeline {
    //指定构建节点
    agent { label "build"}
    stages{
        //下载代码
        stage("CheckOut"){
            steps{
                script{
                    println("CheckOut")
                    checkout([$class: 'GitSCM',
                        branches: [[name: "${params.branchName}"]],
```

```
                extensions: [],
                userRemoteConfigs:[[credentialsId:'gitlab-admin-user',
                url: "${params.srcUrl}"]]])
                sh "ls -l "   //验证
            }
        }
    }
    //构建代码
    stage("Build"){
        steps{
            script{
                println("Build")
                sh "${params.buildShell}"
            }
        }
    }
    //代码扫描
    stage("CodeScan"){
        steps{
            script{
                println("CodeScan")
                withCredentials([usernamePassword(credentialsId:
'702cae96-20f6-42dc-b8c2-d5a5f005e297',
                        passwordVariable: 'PASSWORD',
                        usernameVariable: 'USERNAME')]) {
                    //使用插件方式
withSonarQubeEnv("SonarQube"){
                        sh "sonar-scanner \
                          -Dsonar.login=${USERNAME} \
                          -Dsonar.password=${PASSWORD}"
                    }
                }
            }
        }
    }
}
```

4. 运行 Pipeline

运行 Pipeline 项目后，构建成功页面如图 6-29 所示。

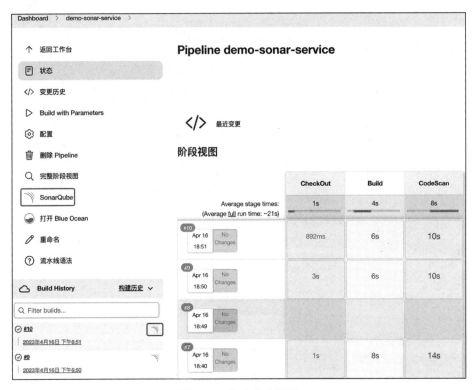

图 6-29 构建成功页面

6.3.4 多分支代码扫描

SonarQube 默认可以支持不同分支的扫描，但是结果仅能展示一份。我们可以通过安装插件实现展示多个分支的扫描结果，代码质量界面如图 6-30 所示。

图 6-30 代码质量界面

1. 安装多分支插件

下载多分支插件的网址为 https://github.com/mc1arke/sonarqube-community-branch-plugin/releases，我们可以下载插件并重启 SonarQube 服务。安装步骤如下。

```
#进入sonarqube容器挂载的目录
cd /data/devops/sonarqube/

#进入插件目录
cd sonarqube_extensions/plugins/

#下载插件
wget https://github.com/mc1arke/sonarqube-community-branch-plugin/releases/download/1.14.0/sonarqube-community-branch-plugin-1.14.0.jar

#切换到配置文件目录
cd /data/devops/sonarqube/sonarqube_conf

#编辑配置文件
vi sonar.properties

#添加以下配置（注意版本号要和插件一致）
sonar.web.javaAdditionalOpts=-javaagent:../extensions/plugins/sonarqube-community-branch-plugin-1.14.0.jar=web
sonar.ce.javaAdditionalOpts=-javaagent:../extensions/plugins/sonarqube-community-branch-plugin-1.14.0.jar=ce

#重启sonarqube
docker restart sonarqube
```

2. 创建分支

多个分支是常见的 Git 开发模式，我们当前的项目只有一个主干 main，下面将基于 main 创建一个 develop 分支，创建分支页面如图 6-31 所示。

图 6-31　创建分支页面

3. 更新 Jenkinsfile

使用多分支插件扫描项目时需要更新扫描参数，完整的 Jenkinsfile 如下。

```
//流水线
pipeline {
    //指定构建节点
    agent { label "build"}
    stages{
        //下载代码
        stage("CheckOut"){
            steps{
                script{
                    println("CheckOut")
                    checkout([$class: 'GitSCM',
                        branches: [[name: "${params.branchName}"]],
                        extensions: [],
                        userRemoteConfigs:[[credentialsId:'gitlab-admin-user',
                        url: "${params.srcUrl}"]]])
                    sh "ls -l "  //验证

                }
            }
        }
        //构建代码
        stage("Build"){
            steps{
                script{
                    println("Build")
                    sh "${params.buildShell}"
                }
            }
        }
        //代码扫描
        stage("CodeScan"){
            steps{
                script{
                    println("CodeScan")
                    withCredentials([usernamePassword(credentialsId:
'702cae96-20f6-42dc-b8c2-d5a5f005e297',
                                passwordVariable: 'PASSWORD',
                                usernameVariable: 'USERNAME')]) {
```

```
                        //使用插件方式
withSonarQubeEnv("SonarQube"){
                        sh "sonar-scanner \
                            -Dsonar.login=${USERNAME} \
                            -Dsonar.password=${PASSWORD}  \
-Dsonar.branch.name=${params.branchName} "
                        }
                    }
                }
            }
        }
    }
}
```

4. 运行 Pipeline

选择 develop 分支，单击"开始构建"按钮，如图 6-32 所示。

图 6-32　Pipeline 构建页面

等待 Pipeline 运行成功后，可以在 SonarQube 中看到 develop 分支的扫描结果，如图 6-33 所示。

图 6-33　SonarQube 项目多分支界面

6.4　本章小结

本章主要讲解了 SonarQube 代码质量平台的配置和集成实践。通过每一步的讲解，读者可深入理解代码质量平台手动代码扫描和与持续集成服务器集成后的配置过程。

第 7 章
制品库平台实战

制品库平台是 DevOps 工具链中的重要组成部分,用于存储、管理、版本控制和应用部署相关的制品,如代码、配置文件、文档、二进制文件等。制品库平台可以帮助开发团队实现制品的集中管理、版本控制、共享和分发,从而提高开发效率和部署效率。

持续集成过程中会产生应用制品,而持续部署过程中会发布应用制品。制品库的最佳实践原则之一是使用同一个制品交付到各个不同的环境。本章笔者将介绍 Nexus Repository 3 的平台实践。

本章内容速览。
- ☑ 制品库平台入门
- ☑ Nexus Repository 实践
- ☑ Nexus Repository 扩展实践
- ☑ 本章小结

7.1 制品库平台入门

制品指的是将源代码构建编译并经过测试和质量检测后生成的二进制文件,如 Java 项目中 War/Jar 格式的包,这些都可以看作是制品。制品库是存放制品的仓库。制品库平台可以集中管理所有的制品,确保制品的版本控制和共享,避免制品混乱和丢失。

组织中存在多个环境,如开发环境、测试环境、预生产环境和生产环境。在没有建设制品平台时,每次持续集成阶段都会重复生成应用制品。搭建制品库平台不仅可以减少重复应用编译的时间,而且可以达到各个环境使用同一个制品交付的原则。

7.1.1 管理规范

在开始引入制品管理的时候,应该根据公司的规模和团队特点定义制品库的管理和使用规范。设置标准化的规范有助于实现持续集成和持续交付，为产品开发创建正确的仓库结构，在产品扩展性方面发挥着至关重要的作用。它不仅可以减少创建管理仓库的开销，还可以帮助团队意识到仓库规范管理的好处，使组织内部各个团队清楚地了解制品的命名规范。

1. 版本号

版本号是软件系统中的一个重要标识符，用于标识软件系统的版本。版本号规范可以帮助开发人员和运维人员更好地理解和管理软件版本，以及在软件版本升级过程中确保版本的一致性和可维护性，如 1.0.0。这种规范通常用于开源软件，其中每个版本都表示一个重大更新，通常使用主版本号表示项目的重大架构变更，次版本号表示较大范围的功能增加和变化，修订版本号表示重大漏洞的修复。

2. 仓库管理

企业中，以开发团队为单位分别创建一个制品库，然后将开发团队中的应用按照不同的目录层级来存储。例如，为开发 1 组创建制品库，名称为 develop01，开发 2 组创建制品库，名称为 develop02，如图 7-1 所示。

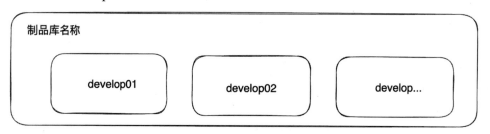

图 7-1　定义制品库名称

制品库中的目录结构可以按照不同的应用名称划分。例如 Java 应用，开发 1 组的应用 app1 的制品存储目录结构为"/所属组-应用名称/应用版本号/应用包名称"，即/develop01-app1/1.1.0/develop01-app1-1.1.0.jar 。

7.1.2　Nexus Repository 3 概述

Nexus Repository 3 是 Nexus 公司的仓库管理平台，它是使用最为广泛的开源仓库管理平台，可以管理整个软件供应链中的组件、二进制文件和构建制品。

Nexus Repository 3 分为社区版和企业版，社区版可以免费且全面地管理二进制文件和制品，企业版具有更多的安全特性。Nexus Repository 3 支持的二进制文件仓库类型如图 7-2

所示。

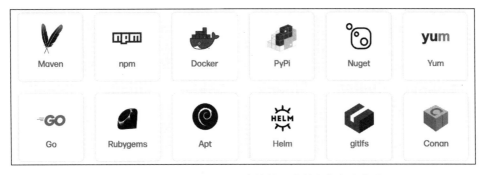

图 7-2 Nexus Repository 3 支持的二进制文件仓库类型

7.2 Nexus Repository 实践

7.2.1 Nexus Repository 3 安装

为了便于测试，准备使用 Docker 容器的方式快速运行 Nexus Repository 3。创建持续化的数据目录，启动容器并挂载本地的持久化数据目录。命令如下。

```
#创建目录
mkdir -p /data/devops/nexus3/data
chmod 777 -R /data/devops/nexus3/

#启动容器
docker run -itd \
-p 8081:8081 \
-v /data/devops6/nexus3/:/nexus-data \
--name nexus3 sonatype/nexus3:3.53.0
```

通过 docker logs -f nexus3 命令查看容器的日志，并确定 Nexus 是否已准备就绪。Nexus Repository 3 的默认端口是 8081，现在转到浏览器并打开网址 http://主机 IP:8081。Nexus Repository 3 平台首页如图 7-3 所示。

安装完成后，默认的 admin 账号和密码被存储在数据目录，获取初始化密码的命令如下。

```
docker exec -i nexus3 cat /nexus-data/admin.password
```

单击 Sign in 按钮，使用初始化密码登入 Nexus Repository 3，如图 7-4 所示。

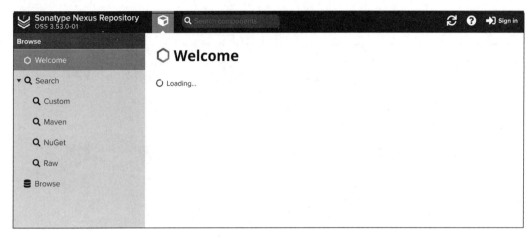

图 7-3　Nexus Repository 3 平台首页

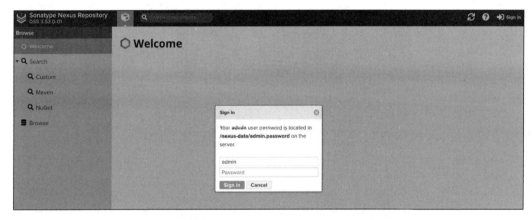

图 7-4　登入 Nexus Repository 3

7.2.2　搭建 Maven 私服仓库

通常情况下，开发人员在进行代码开发时会使用一些包管理工具，如 Maven、Gradle，这些都是常见项目的编译构建工具。我们可以将这些工具理解为命令行工具。

使用私服，就是在企业内部建立中央存储仓库。例如，在公司内部通过 Nexus Repository 3 创建一个代理仓库，将公网仓库中的 Maven 包代理到内网仓库中。这样就可以直接访问内网的私服下载、构建依赖包。内网的速度要比公网快，这会直接加快管道的构建速度。

代理仓库不会把公网仓库中的所有包下载到本地，而是按需缓存。例如，此时需要使用 aa 这个包，如果代理仓库中没有，则请求外部服务器下载这个包并进行缓存，当第二次访问时，即可直接访问代理仓库。

1. 创建代理仓库

导航进入 Nexus Repository 3 的管理页面，单击 Create repository 按钮创建仓库。Nexus Repository 3 创建仓库页面如图 7-5 所示。

图 7-5　Nexus Repository 3 创建仓库页面

2. 填写仓库信息

创建仓库时选择 Proxy 类型，即代理仓库。可以自定义仓库名称，在 Remote storage 文本框中填写远程要代理的仓库日志。图 7-6 所示为代理阿里云仓库。

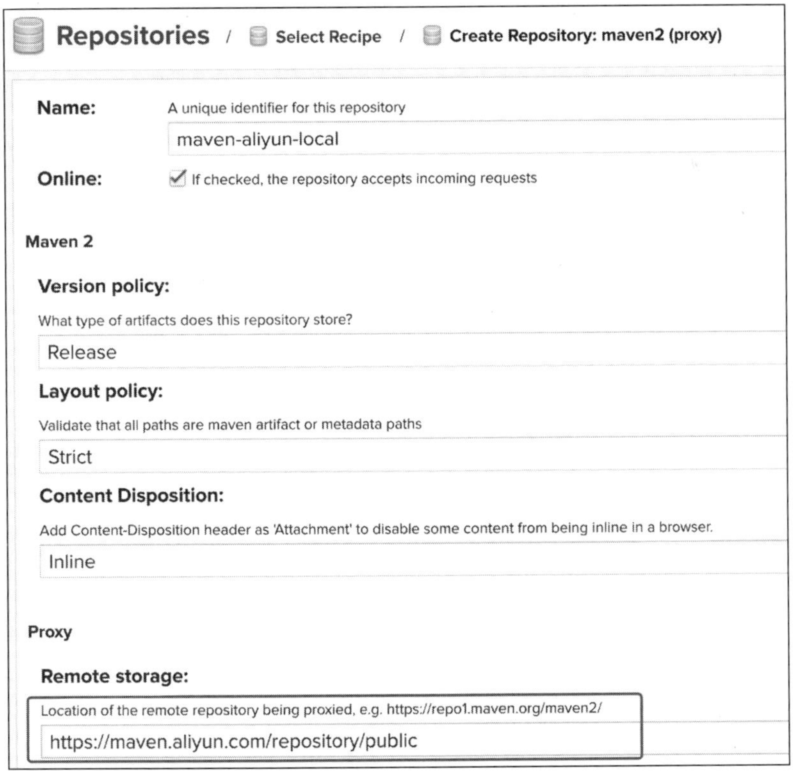

图 7-6　代理阿里云仓库

3. 修改 Maven 配置

填写好仓库参数,创建仓库后会有对应的仓库地址,如图 7-7 所示。

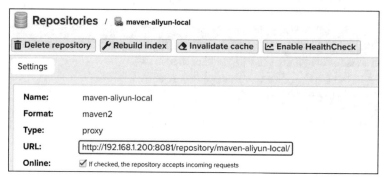

图 7-7 代理仓库地址

Maven 客户端要想使用私服,需要编辑配置文件。编辑 settings.xml 文件并添加 mirror 配置,配置如下。

```
<mirror>
    <id>aliyunmaven</id>
    <mirrorOf>*</mirrorOf>
    <name>代理仓库</name>
    <url>http://192.168.1.200:8081/repository/maven-aliyun-local/</url>
</mirror>
```

4. 运行 Maven 构建测试

在项目源码目录中,通过 mvn clean package 命令进行编译构建。如果提示的日志中对应的是所配置的私服仓库地址,说明配置已经生效。日志如下。

```
mvn clean package
[INFO] Scanning for projects...
Downloading from aliyunmaven: http://192.168.1.200:8081/repository/maven-aliyun-local/org/springframework/boot/spring-boot-starter-parent/2.7.10/spring-boot-starter-parent-2.7.10.pom
Downloaded from aliyunmaven: http://192.168.1.200:8081/repository/maven-aliyun-local/org/springframework/boot/spring-boot-starter-parent/2.7.10/spring-boot-starter-parent-2.7.10.pom (9.2 kB at 75 kB/s)
```

7.2.3 搭建 Maven 本地仓库

本地仓库一般用于存储企业项目组中自己开发的包。以 Maven 为例,分为 Release 类型仓库（存放稳定版制品）和 Snapshot 类型仓库（存放开发版制品）两种。仓库的版本策略可以在创建仓库时选择，如图 7-8 所示。

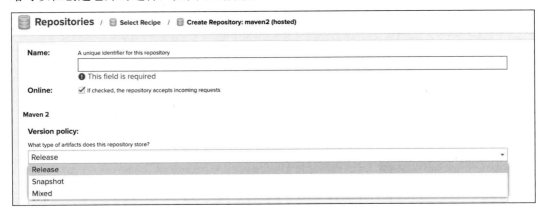

图 7-8　在创建仓库中选择版本策略

在企业中，通常会分别创建不同版本的仓库。Release 类型的仓库只能存放 Release 版本的包。将 Snapshot 类型的包上传到 Release 类型的仓库中时会提示错误。Release 类型和 Snapshot 类型的仓库如图 7-9 所示。

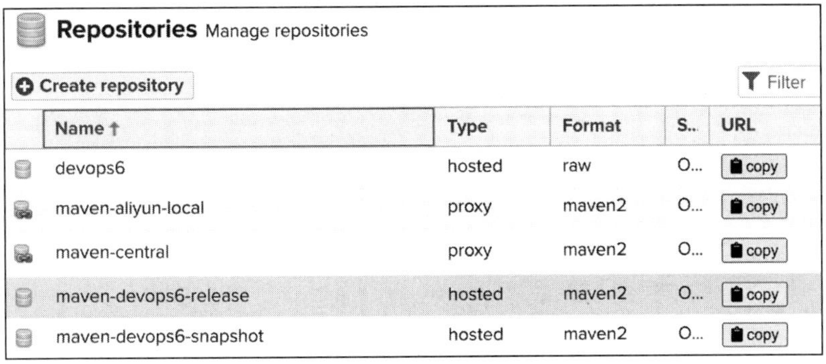

图 7-9　Release 类型和 Snapshot 类型的仓库

7.2.4　制品上传方式

1. Web 页面上传

Nexus Repository 3 提供了通过 Web 页面上传制品的途径。单击左侧菜单的 Upload，然后选择要上传的目标仓库，如图 7-10 所示。

选中仓库之后需要选择包路径和目标目录。确认参数无误后，单击 Upload 按钮完成上传，包的参数信息如图 7-11 所示。

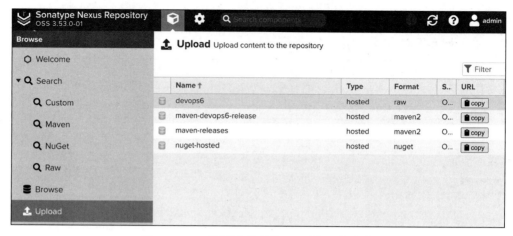

图 7-10　Web 页面上传制品——选择目标仓库

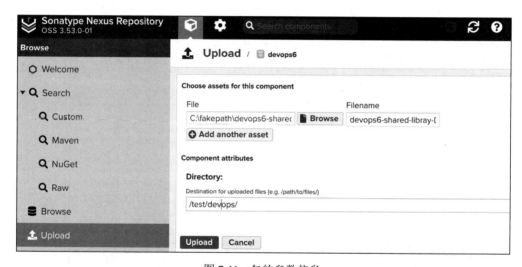

图 7-11　包的参数信息

2. Maven 命令上传

在 Nexus Repository 3 中，Maven 类型的仓库支持通过 Maven 命令行上传制品。

首先，需要把与私服通信的账号和密码存储到配置文件中。编辑 Maven 的配置文件，在 settings.xml 中添加仓库的认证信息。配置如下。

```
<server>
 <id>maven-devops6-release</id>
 <username>admin</username>
 <password>admin123</password>
</server>
```

注意，使用 mvn deploy 发布时，-DrepositoryId 参数的值要与上面配置文件中的<server>标签中的<id>一致。不然会出现错误代码为 401 的用户认证失败问题。上传命令如下。

```
mvn deploy:deploy-file
-DgroupId=com.demo                              #pom 中的 groupId
-DartifactId=demo                               #pom 中的 artifactId
-Dversion=1.1.2                                 #pom 中的版本号 version
-Dpackaging=jar                                 #pom 中打包方式
-Dfile=target/demo-0.0.1-SNAPSHOT.jar           #本地文件
-Durl=http://192.168.1.200:8081/repository/maven-devops6-release/#仓库 url
-DrepositoryId=maven-devops6-release            #对应的是 setting.xml（认证）
```

上传日志中可以看到制品上传的整个过程，如图 7-12 所示。

```
[root@devops-nuc-service devops6-maven-service]# mvn deploy:deploy-file -DgroupId=com.demo -DartifactId=demo -Dversion=1.1.2 -Dpackaging
=jar -Dfile=target/demo-0.0.1-SNAPSHOT.jar -Durl=http://192.168.1.200:8081/repository/maven-devops6-release/ -DrepositoryId=maven-devops
6-release
[INFO] Scanning for projects...
[INFO]
[INFO] ------------------------< com.example:demo >------------------------
[INFO] Building demo 0.0.1-SNAPSHOT
[INFO]   from pom.xml
[INFO] --------------------------------[ jar ]---------------------------------
[INFO]
[INFO] --- deploy:2.8.2:deploy-file (default-cli) @ demo ---
[          ]
Uploading to maven-devops6-release: http://192.168.1.200:8081/repository/maven-devops6-release/com/demo/demo/1.1.2/demo-1.1.2.jar
Uploaded to maven-devops6-release: http://192.168.1.200:8081/repository/maven-devops6-release/com/demo/demo/1.1.2/demo-1.1.2.jar (18 MB
 at 30 MB/s)
Uploading to maven-devops6-release: http://192.168.1.200:8081/repository/maven-devops6-release/com/demo/demo/1.1.2/demo-1.1.2.pom
Uploaded to maven-devops6-release: http://192.168.1.200:8081/repository/maven-devops6-release/com/demo/demo/1.1.2/demo-1.1.2.pom (386 B
 at 12 kB/s)
Downloading from maven-devops6-release: http://192.168.1.200:8081/repository/maven-devops6-release/com/demo/demo/maven-metadata.xml
Downloaded from maven-devops6-release: http://192.168.1.200:8081/repository/maven-devops6-release/com/demo/demo/maven-metadata.xml (292
 B at 19 kB/s)
Uploading to maven-devops6-release: http://192.168.1.200:8081/repository/maven-devops6-release/com/demo/demo/maven-metadata.xml
Uploaded to maven-devops6-release: http://192.168.1.200:8081/repository/maven-devops6-release/com/demo/demo/maven-metadata.xml (323 B at
 8.7 kB/s)
[INFO] ------------------------------------------------------------------------
[INFO] BUILD SUCCESS
[INFO] ------------------------------------------------------------------------
[INFO] Total time:  1.241 s
[INFO] Finished at: 2023-05-07T09:47:06+08:00
[INFO] ------------------------------------------------------------------------
```

图 7-12　制品上传过程

命令行提示成功后，可以进入 Nexus Repository 3 平台验证。如图 7-13 所示，制品已经上传成功。

上面我们通过命令行的-D 选项来传递制品的参数，我们也可以通过直接读取 pom.xml 文件中定义的参数上传制品。即添加-DpomFile 参数指定 pom.xml 文件的路径，命令如下。

```
mvn deploy:deploy-file \
-DpomFile=pom.xml \
-Dfile=target/demo-0.0.1-SNAPSHOT.jar \
-Durl=http://192.168.1.200:8081/repository/maven-devops6-release/ \
-DrepositoryId=maven-devops6-release
```

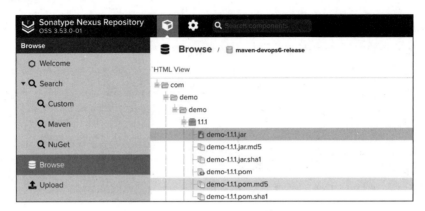

图 7-13　验证制品上传成功

7.3　Nexus Repository 扩展实践

7.3.1　调试 REST API

进入设置页面，选择 System→API，即可进入 API 调试页面，如图 7-14 所示。

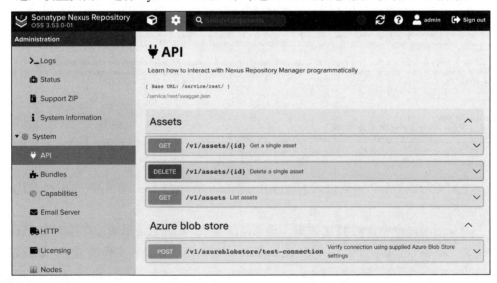

图 7-14　Nexus Repository 3 API 页面

下面以上传制品的接口/v1/components 为例进行调试（其他接口的调试过程与此接口类似）。单击 Try it out 按钮开启调试，如图 7-15 所示。

第 7 章 制品库平台实战

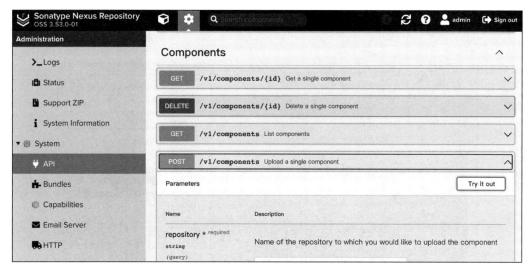

图 7-15 调试上传制品接口

这个接口可以上传任何类型的制品，我们以 Maven 类型的制品为例。首先填写 repository 参数的仓库名称，如图 7-16 所示。

然后填写 Maven 制品的坐标信息 groupId、artifactId、version，如图 7-17 所示。

继续填写 packaging、asset1、extension 等信息，如图 7-18 所示。

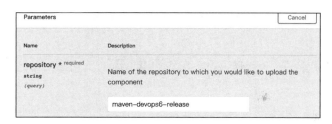

图 7-16 填写仓库名称

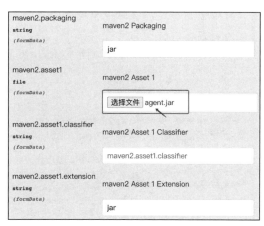

图 7-17 填写制品坐标信息　　图 7-18 填写制品类型和路径

填写制品信息完成后,单击 Execute 按钮执行操作,状态码 204 表示成功,上传制品如图 7-19 所示。

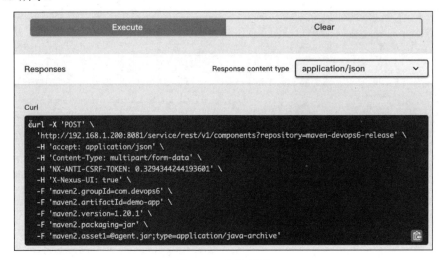

图 7-19　上传制品

7.3.2　上传 Raw 类型制品

如果项目编译生成的制品需要被依赖,可以通过创建 Maven 类型的仓库进行管理;如果不需要被其他项目依赖,则可以创建 Raw 类型的仓库管理。Raw 类型的仓库即通用类型,可以理解为一块通用的文件存储,支持各种格式(如 PNG、tar.gz、zip、jar)的制品等。

Nexus Repository 3 支持很多种类型的制品仓库,这里笔者把通用制品类型的上传 API 接口和对应的 curl 测试命令整理如下。

```
##上传 PNG 格式文件
curl -X POST "http://192.168.1.200:8081/service/rest/v1/components?repository=myrepo" \
-H "accept: application/json" \
-H "Content-Type: multipart/form-data" \
-F "raw.directory=/tmp" \
-F "raw.asset1=@test.png;type=image/png" \
-F "raw.asset1.filename=test.png"

##上传 tar.gz 和 zip 格式文件
curl -X POST "http://192.168.1.200:8081/service/rest/v1/components?repository=myrepo" \
-H "accept: application/json" \
```

```
-H "Content-Type: multipart/form-data" \
-F "raw.directory=/tmp" \
-F "raw.asset1=@nexus-3.30.0-01-unix.tar.gz;type=application/x-gzip" \
-F "raw.asset1.filename=aaa.tar.gz"

##上传jar格式文件
curl -X POST "http://192.168.1.200:8081/service/rest/v1/components?repository=myrepo" \
-H "accept: application/json" \
-H "Content-Type: multipart/form-data" \
-F "raw.directory=/tmp" \
-F "raw.asset1=@aopalliance-1.0.jar;type=application/java-archive" \
-F "raw.asset1.filename=aopalliance-1.0.jar"
```

7.3.3 Jenkins 插件上传制品

进入 Jenkins 插件管理，搜索并安装 Nexus Artifact Uploader 插件，如图 7-20 所示。

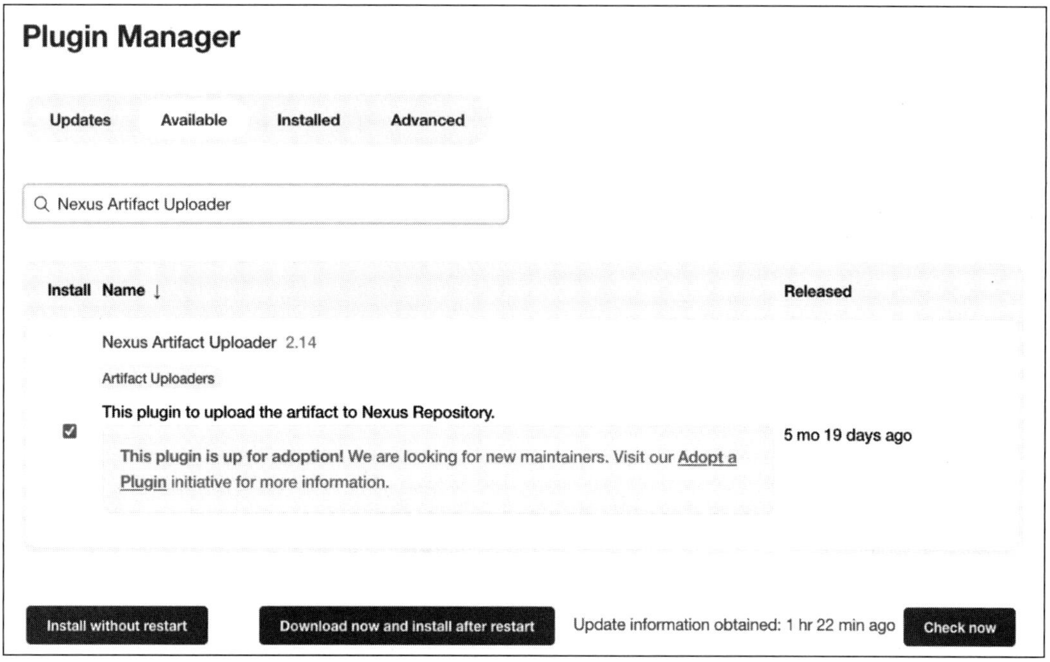

图 7-20　安装 Nexus Artifact Uploader 插件

使用片段生成器生成流水线代码，如图 7-21 所示。

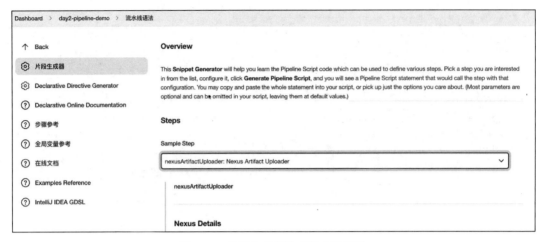

图 7-21 片段生成器生成流水线代码

最后,笔者整理了一个完整的 Jenkins Pipeline 代码,具体如下。

```
pipeline {
    agent { label "build"}
    options {
        skipDefaultCheckout true
    }
    stages{
        stage("CheckOut"){
            steps{
                script{
                    println("CheckOut")
                }
            }
        }
        stage("Build"){
            steps{
                script{
                    println("Build")
                }
            }
        }
        stage("PushArtifact"){
            steps{
                script{
                    println("PushArtifact")
                    //上传制品
                    PushArtifactByPlugin()
```

```
                }
            }
        }
    }
}
//使用插件上传
def PushArtifactByPlugin(){
    nexusArtifactUploader artifacts: [[artifactId: 'demo-app',
                                       classifier: '',
                                       file: 'target/demo-0.0.1-SNAPSHOT.jar',
                                       type: 'jar']],
                          credentialsId:'71cbafcb-6447-4213-b214-46535a2ef733',
                          groupId: 'com.devops6',
                          nexusUrl: '192.168.1.200:8081',
                          nexusVersion: 'nexus3',
                          protocol: 'http',
                          repository: 'maven-devops6-release',
                          version: '1.1.1'
}
```

7.4 本章小结

本章，笔者分别从安装、搭建私服仓库，搭建本地仓库，制品上传，调试 API，Jenkins 集成等方面讲解了 Nexus Repository 3 制品管理平台实践。读者通过建设制品库平台可以提高持续集成和持续部署的速度和质量。

第 8 章
云主机环境持续部署

本章主要讲述传统应用运行环境中的持续部署实践。在企业私有云或者公有云环境中通常使用云主机作为应用服务的运行环境，笔者将通过 Jenkins 集成 Ansible 实现从持续集成到持续部署的整个过程。

本章内容速览。
- ☑ 项目准备工作
- ☑ 持续集成实践
- ☑ 持续部署实践
- ☑ 本章小结

8.1 项目准备工作

根据笔者实施 DevOps 项目流程改造的经验，建议先了解当前项目团队中的开发流程，发现可能存在的流程问题并制定一些发布标准后再进行实际流水线的开发和落地。这将有助于形成 DevOps 试点标准项目模板，然后根据标准化实现流程工具和代码的复用。

流水线的实践方式各式各样，适合自身团队的才是最有价值的。所以不要照搬其他团队或者公司、组织的模式，那些模式未必适合自己。

8.1.1 分支策略

分支策略分为基于特性分支开发和基于版本分支发布：基于主干分支创建特性分支和版本分支，特性分支开发完成并经过验证后合并到版本分支；基于版本分支进行持续集成流水线生成制品，然后通过持续部署流水线将制品发布到各个环境，分支策略如图 8-1 所示。

选择版本分支发布是为避免对主干分支污染。如果发布出现问题，可基于版本分支进行代码修复，然后继续持续集成和持续部署流水线。发布生产完成后，将版本分支合并到主干分支并基于主干分支创建一个版本标签。

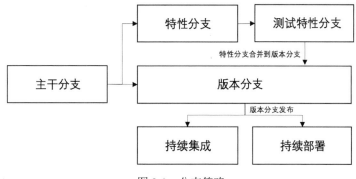

图 8-1 分支策略

8.1.2 环境准备

在企业中基于不同的业务，应用可能部署在私有云或者公有云。这里我们通过两台云主机进行模拟，本地环境可以采用 VirtualBox 创建两台虚拟机，如图 8-2 所示。

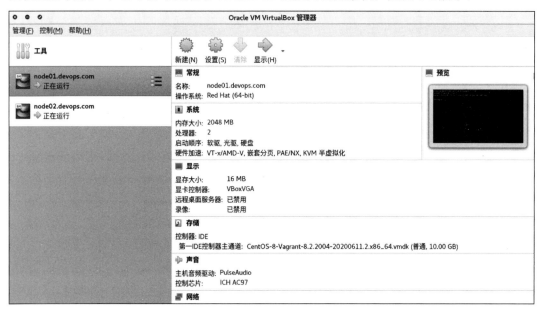

图 8-2 VirtualBox 虚拟机配置

笔者本地机器的信息如下。

- ☑ 主机名称：node01.devops.com，node02.devops.com。
- ☑ 配置：2 核 CPU，2GB 内存。
- ☑ IP 地址：192.168.1.121/24，192.168.1.122/24。
- ☑ 操作系统：CentOS 8。

由于笔者演示的是 Java 项目的发布，因此需要在两台机器上安装好 JDK 环境。命令如下。

```
#下载 JDK11
wget https://mirrors.tuna.tsinghua.edu.cn/Adoptium/11/jdk/x64/linux/OpenJDK11U-jdk_x64_linux_hotspot_11.0.19_7.tar.gz

#解压
tar zxf OpenJDK11U-jdk_x64_linux_hotspot_11.0.19_7.tar.gz -C /usr/local

#配置环境变量，编辑/etc/profile 文件
export JAVA_HOME=/usr/local/jdk11
export PATH=$JAVA_HOME/bin:$PATH

#生效环境变量
source/etc/profile

#验证版本，出现版本信息即成功
java -version
```

8.1.3　Ansible 配置

Ansible 是一个 IT 自动化工具，简单易用。使用 OpenSSH 进行传输，可以配置系统、部署软件以及编排更高级的运维任务。Jenkins 流水线中需要使用 Ansible 命令行管理机器，我们可以将 Ansible 安装到 Jenkins Agent 节点。安装命令如下。

```
#安装 Ansible
dnf install https://dl.fedoraproject.org/pub/epel/epel-release-latest-8.noarch.rpm -y
dnf install ansible
#验证版本
ansible --version
```

Ansible 如何管理主机呢？Ansible 读取主机清单文件，对主机进行分类，配置 Ansible 管理的主机，默认文件为/etc/ansible/hosts。将两台云主机加入清单文件。内容如下。

```
[webserver]
192.168.1.121
192.168.1.122
```

配置 Jenkins Agent 节点到两台云主机的 SSH 免密认证。命令如下。

```
#配置免密
ssh-copy-id root@192.168.1.121
ssh-copy-id root@192.168.1.122
```

运行 Ansible 命令进行测试。测试命令如下。

```
#测试连通性
ansible all -m ping -u root
#复制文件
ansible webserver -m copy -a "src=/etc/passwd dest=/opt/passwd"
```

8.1.4 Pipeline 设计

1. 持续集成流程

持续集成的流程应该包含下载项目代码、编译构建、通过 SonarQube 进行质量检查、检查通过后上传制品到 Nexus 仓库，持续集成流水线设计如图 8-3 所示。

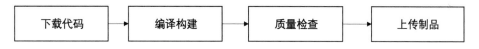

图 8-3 持续集成流水线设计

2. 持续部署流程

持续部署的流程应该包含下载制品、应用发布、人工验证、应用回滚，如图 8-4 所示。

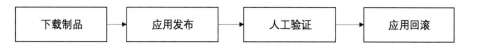

图 8-4 持续部署流水线设计

3. 配置共享库

Jenkins 流水线即代码，创建 Jenkins 共享库来配置流水线，实现流水线代码模块化以及复用。在 Jenkins 系统设置中添加共享库，填写共享库的名称、默认分支版本、远程存储库地址、存储库认证凭据，共享库配置如图 8-5 所示。

图 8-5　共享库配置

8.2　持续集成实践

8.2.1　准备工作

1．创建 Pipeline

下面创建一个持续集成流水线作业，如果将这个作业作为模板进行复用，这将便于后期若干流水线作业的创建。这里笔者将一些通用的参数定义在 Jenkins 作业页面中展示，以便于参数的可视化。持续集成流水线参数如图 8-6 所示。

参数解释如下。
- ☑　srcUrl：项目代码库地址。
- ☑　buildShell：项目构建命令。
- ☑　branchName：项目分支名称。

第 8 章　云主机环境持续部署

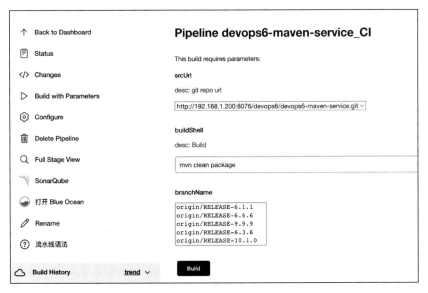

图 8-6　持续集成流水线参数

> **提示：**
> 这里配置的 branchName 参数是 Git Parameter 类型。需要安装 Git Parameter 插件并重启 Jenkins。使用 Git Parameter 参数时一定要配置好 Use repository 参数，不然会出现无法加载分支的问题。Git Parameter 参数配置如图 8-7 所示。

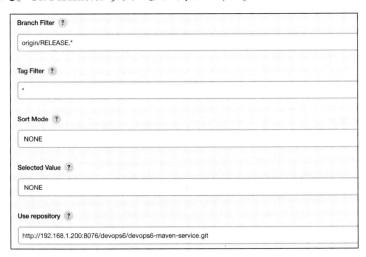

图 8-7　Git Parameter 参数配置

2. 配置 Jenkinsfile

选择加载共享库中的 Jenkinsfile。笔者将 Jenkinsfile 和共享库代码放到一起，持续集成流水线用的 Jenkinsfile 名称是 ci.jenkinsfile。SCM 选项用于选择仓库类型，这里选择 Git 类型。填写文件保存的远程 Git 仓库地址、认证凭据、分支名称、脚本路径等，Jenkinsfile 脚本配置如图 8-8 所示。

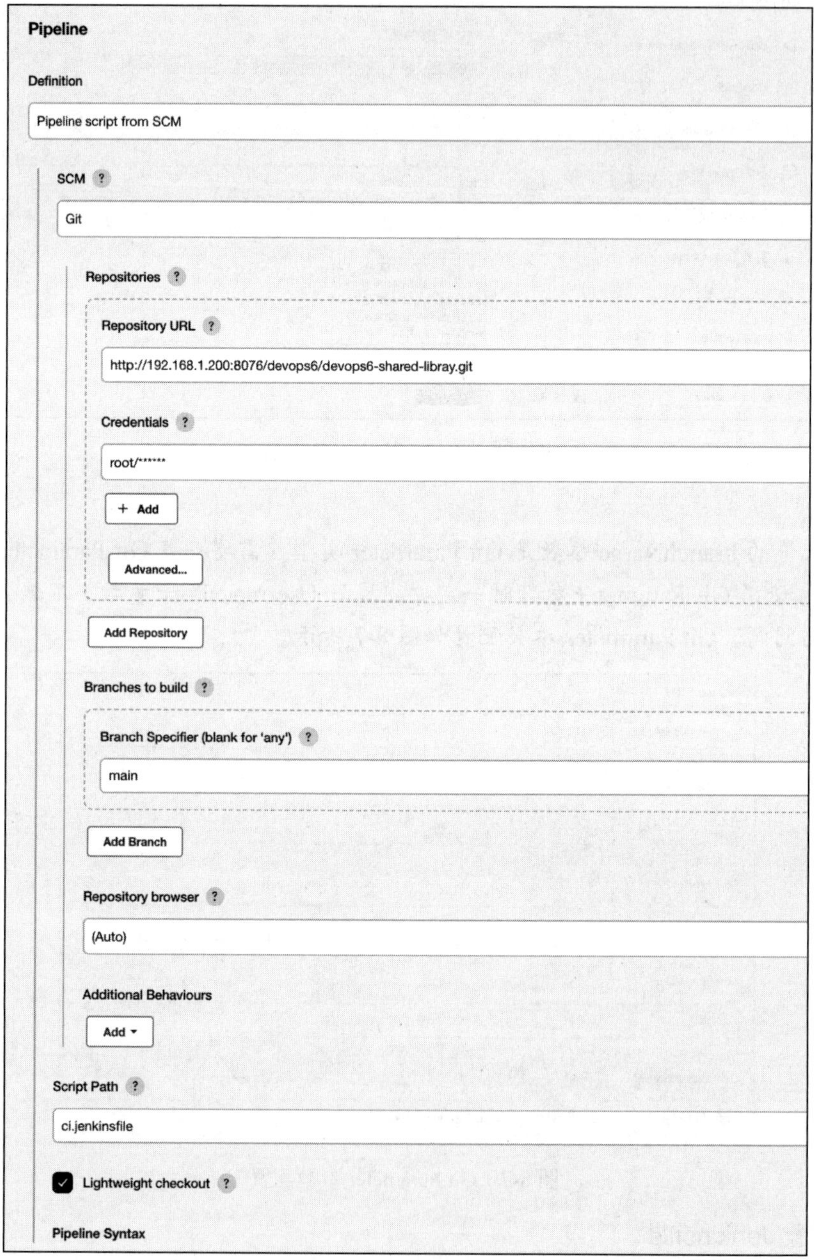

图 8-8　Jenkinsfile 脚本配置

3. 加载共享库

笔者在共享库中编写了一些关于持续集成的模块，在 Jenkinsfile 中加载共享库并导入共享库中的模块。代码如下。

```
//加载共享库
@Library("devops06@main") _

//导入模块
def build = new org.devops.Build()
def sonar = new org.devops.Sonar()
def artifact = new org.devops.Artifact()
def checkout = new org.devops.Checkout()
```

8.2.2　设置 Pipeline

设置 Pipeline 运行的构建节点、运行时选项、持续集成的多个阶段。持续集成阶段分为 CheckOut（下载代码）、Build（编译构建）、CodeScan（质量检查）、PushArtifact（上传制品）。代码如下。

```
//使用git参数需要格式化
//删除分支前面的"origin/"字符
env.branchName = "${env.branchName}" - "origin/"
println(env.branchName)

//持续集成流水线
pipeline {
    agent { label "build"}           //运行在build标签的节点
    options {
        skipDefaultCheckout true     //跳过默认的代码检出
    }
    stages{
        stage("CheckOut"){           //下载代码
            steps{
                script{
                    println("CheckOut")
                }
            }
        }
        stage("Build"){              //编译构建
            steps{
                script{
```

```
                println("Build")
            }
        }
    }
    stage("CodeScan"){              //质量检查
        steps{
            script{
                println("CodeScan")
            }
        }
    }
    stage("PushArtifact"){           //上传制品
        steps{
            script{
              println("PushArtifact")
            }
        }
    }
}
```

1. 下载代码

在共享库 org/devops/Build.groovy 中编写 CheckOut()函数。该函数调用 Jenkins 内置语法 checkout 语句来下载项目的代码并运行 Shell 命令验证代码的下载。代码如下。

```
//CheckOut
def CheckOut(){
   println("CheckOut")
   checkout([$class: 'GitSCM',
       branches: [[name: "${env.branchName}"]],
       extensions: [],
       userRemoteConfigs: [[credentialsId: 'gitlab-admin',
       url: "${params.srcUrl}"]]])
   sh "ls -l "                            //验证
}
```

在 Pipeline 的 CheckOut 阶段可以直接调用 Build.groovy 文件中定义的 CheckOut()函数。代码如下。

```
stage("CheckOut"){
    steps{
        script{
          build.CheckOut()
```

```
        }
    }
}
```

下载代码阶段的日志如图 8-9 所示。

```
[Pipeline] {
[Pipeline] stage
[Pipeline] { (CheckOut)
[Pipeline] script
[Pipeline] {
[Pipeline] echo
CheckOut
[Pipeline] checkout
The recommended git tool is: NONE
using credential gitlab-admin
Fetching changes from the remote Git repository
Checking out Revision 58766f18f72175aec1e048cccea48e47a06343ea (origin/RELEASE-10.1.0)
Commit message: "Merge branch 'feature-dev-666' into 'RELEASE-10.1.0'"
First time build. Skipping changelog.
[Pipeline] sh
+ ls -l
总用量 28
-rwxr-xr-x 1 root root 10284 5月  20 10:56 mvnw
-rw-r--r-- 1 root root  6734 5月  20 10:56 mvnw.cmd
-rw-r--r-- 1 root root  1246 5月  20 10:56 pom.xml
-rw-r--r-- 1 root root   580 5月  20 10:56 sonar-project.properties
drwxr-xr-x 4 root root    30 5月  20 10:56 src
drwxr-xr-x 9 root root   215 5月  20 10:58 target
[Pipeline] }
```

图 8-9 下载代码阶段日志

2．编译构建

在共享库 org/devops/Build.groovy 中编写 Build() 函数。该函数调用 Shell 命令编译项目构建。例如，此处将会加载在 Jenkins 页面中定义的 buildShell 参数的值，即运行 mvn clean package 命令进行项目编译构建。代码如下。

```
//编译构建
def Build(){
    println("Build")
    sh "${params.buildShell}"  //构建命令
}
```

在 Pipeline 的 Build 阶段可以直接调用 Build.groovy 文件中定义的 Build() 函数。代码如下。

```
stage("Build"){
    steps{
        script{
            build.Build()
```

```
        }
    }
}
```

编译构建阶段的日志如图 8-10 所示。

```
  .   ____          _            __ _ _
 /\\ / ___'_ __ _ _(_)_ __  __ _ \ \ \ \
( ( )\___ | '_ | '_| | '_ \/ _` | \ \ \ \
 \\/  ___)| |_)| | | | | || (_| |  ) ) ) )
  '  |____| .__|_| |_|_| |_\__, | / / / /
 =========|_|==============|___/=/_/_/_/
 :: Spring Boot ::                (v2.7.10)

2023-05-20 11:07:00.814  INFO 34556 --- [           main] com.example.demo.DemoApplicationTests    : Starting DemoApplicationTests
nuc-service with PID 34556 (started by root in /opt/jenkinsagent/workspace/devops6-maven-service_CI)
2023-05-20 11:07:00.815  INFO 34556 --- [           main] com.example.demo.DemoApplicationTests    : No active profile set, falling
"default"
2023-05-20 11:07:01.983  INFO 34556 --- [           main] com.example.demo.DemoApplicationTests    : Started DemoApplicationTests i
2.287)
[INFO] Tests run: 1, Failures: 0, Errors: 0, Skipped: 0, Time elapsed: 2.066 s - in com.example.demo.DemoApplicationTests
[INFO]
[INFO] Results:
[INFO]
[INFO] Tests run: 1, Failures: 0, Errors: 0, Skipped: 0
[INFO]
[INFO]
[INFO] --- jar:3.2.2:jar (default-jar) @ demo ---
[INFO] Building jar: /opt/jenkinsagent/workspace/devops6-maven-service_CI/target/demo-2.1.1.jar
[INFO]
[INFO] --- spring-boot:2.7.10:repackage (repackage) @ demo ---
[INFO] Replacing main artifact with repackaged archive
[INFO] ------------------------------------------------------------------------
[INFO] BUILD SUCCESS
[INFO] ------------------------------------------------------------------------
[INFO] Total time:  5.183 s
[INFO] Finished at: 2023-05-20T11:07:03+08:00
[INFO] ------------------------------------------------------------------------
```

图 8-10　编译构建阶段日志

3．质量检查

在共享库 org/devops/Sonar.groovy 中编写 SonarScannerByPlugin()函数。该函数封装了 SonarQube 插件提供的 withSonarQubeEnv 语句进行代码扫描。由于代码扫描需要用到 SonarQube 用户认证，读者需要将其保存在 Jenkins 凭据中。代码如下。

```
//SonarScannerByPlugin
def SonarScannerByPlugin(){
    withSonarQubeEnv(credentialsId: 'fdf4362a-69e7-4014-8fa7-80b1ba268588') {
        withCredentials([usernamePassword(credentialsId:
'9ff42b72-597e-49dd-a62f-2553d48304fc',
                                           passwordVariable: 'SONAR_PASSWD',
                                           usernameVariable: 'SONAR_USER')]) {
```

```
            sh """
                sonar-scanner \
                -Dsonar.login=${SONAR_USER} \
                -Dsonar.password=${SONAR_PASSWD} \
                -Dsonar.host,url=http://192.168.1.200:9000 \
                -Dsonar.branch.name=${env.branchName}
                """
        }
    }
}
```

在 Pipeline 的 CodeScan 阶段可以调用 Sonar.groovy 文件中的 SonarScannerByPlugin() 函数。代码如下：

```
stage("CodeScan"){
    steps{
        script{
            sonar.SonarScannerByPlugin()
        }
    }
}
```

质量检查阶段日志如图 8-11 所示。

```
INFO: ------------- Run sensors on project
INFO: Sensor Analysis Warnings import [csharp]
INFO: Sensor Analysis Warnings import [csharp] (done) | time=0ms
INFO: Sensor Zero Coverage Sensor
INFO: Sensor Zero Coverage Sensor (done) | time=5ms
INFO: Sensor Java CPD Block Indexer
INFO: Sensor Java CPD Block Indexer (done) | time=12ms
INFO: SCM Publisher SCM provider for this project is: git
INFO: SCM Publisher 1 source file to be analyzed
INFO: SCM Publisher 1/1 source file have been analyzed (done) | time=143ms
INFO: CPD Executor 3 files had no CPD blocks
INFO: CPD Executor Calculating CPD for 0 files
INFO: CPD Executor CPD calculation finished (done) | time=0ms
INFO: Load New Code definition
INFO: Load New Code definition (done) | time=175ms
INFO: Analysis report generated in 237ms, dir size=129.5 kB
INFO: Analysis report compressed in 13ms, zip size=22.5 kB
INFO: Analysis report uploaded in 177ms
INFO: ANALYSIS SUCCESSFUL, you can find the results at: http://192.168.1.200:9000/dashboard?id=devops6-maven-service&branch=RELEASE-10.1.0
INFO: Note that you will be able to access the updated dashboard once the server has processed the submitted analysis report
INFO: More about the report processing at http://192.168.1.200:9000/api/ce/task?id=AYg3ICZukBN0n0M7GODA
INFO: Analysis total time: 8.959 s
INFO: ------------------------------------------------------------
INFO: EXECUTION SUCCESS
INFO: ------------------------------------------------------------
INFO: Total time: 10.222s
INFO: Final Memory: 22M/88M
INFO: ------------------------------------------------------------
```

图 8-11 质量检查阶段日志

代码质量结果如图 8-12 所示。

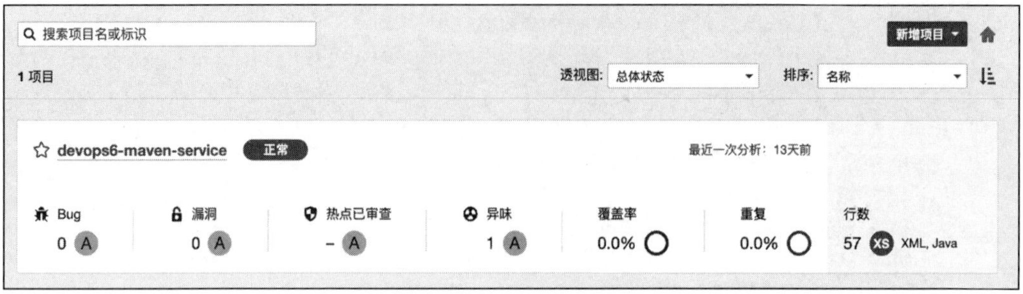

图 8-12　代码质量结果

4．上传制品

在共享库 org/devops/Artifact.groovy 中编写 PushNexusArtifact()函数。该函数封装了上传构建后生成的制品到远程 Nexus 制品仓库的逻辑。代码如下。

```
//Push artifact
def PushNexusArtifact(repoId, targetDir, pkgPath, sourcePkgName, targetPkgName){
    //Nexus API
    withCredentials([usernamePassword(credentialsId: '71cbafcb-6447-4213-b214-46535a2ef733', \
                                passwordVariable: 'PASSWD',
                                usernameVariable: 'USERNAME')]) {
        sh """
            curl -X 'POST' \
            "http://192.168.1.200:8081/service/rest/v1/components?repository=${repoId}" \
            -H 'accept: application/json' \
            -H 'Content-Type: multipart/form-data' \
            -F "raw.directory=${targetDir}" \
            -F "raw.asset1=@${pkgPath}/${sourcePkgName};type=application/java-archive" \
            -F "raw.asset1.filename=${targetPkgName}" \
            -u ${USERNAME}:${PASSWD}
        """
    }
}
```

该函数具有 4 个参数，即 repoId（仓库名称）、targetDir（目标目录）、sourcePkgName（源包名称）、targetPkgName（上传后包名称）。由于上传制品库需要用到身份验证，读者需要将其保存在 Jenkins 凭据中（此处参考 8.3.2 节）。

在 Pipeline 的 PushArtifact 阶段可以调用 Artifact.groovy 文件中的 PushNexusArtifact() 函数。代码如下。

```
stage("PushArtifact"){
    steps{
        script{
            //PushArtifactByPlugin()
            //PushArtifactByPluginPOM()
            //init package info
            appName = "${JOB_NAME}".split('_')[0] //devops6-maven-service
            repoName = appName.split('-')[0]    //devops6
            //获取 commitID
            commitID = checkout.GetCommitID()
            println(commitID)
            appVersion = "${env.branchName}".split("-")[-1]
            appVersion = "${appVersion}-${commitID}"
            targetDir="${appName}/${appVersion}"

            //通过 pom 文件获取包名称
            POM = readMavenPom file: 'pom.xml'
            env.artifactId = "${POM.artifactId}"
            env.packaging = "${POM.packaging}"
            env.groupId = "${POM.groupId}"
            env.art_version = "${POM.version}"
            sourcePkgName = "${env.artifactId}-${env.art_version}.${env.packaging}"
            pkgPath = "target"
            targetPkgName = "${appName}-${appVersion}.${env.packaging}"
            //上传制品
            artifact.PushNexusArtifact(repoName, targetDir, pkgPath, sourcePkgName,targetPkgName)

        }
    }
}
```

上传制品阶段存在对一些数据的解析，根据 Jenkins 作业名称解析获取应用的名称和对应的制品库名称。然后调用 GetCommitID()函数运行 Shell 命令获取提交 ID。上传制品的命名格式是"应用名称-应用版本"，应用版本由版本分支中的版本号和提交 ID 组成。

原制品名称和目标制品名称是不一致的。在 Maven 项目构建编译时根据 pom.xml 加载配置。此处通过 readMavenPom 方法读取 pom.xml 文件，然后获取由 artifactId、version、packaging 组成的原始生成的制品名称。

上传制品阶段日志如图 8-13 所示。

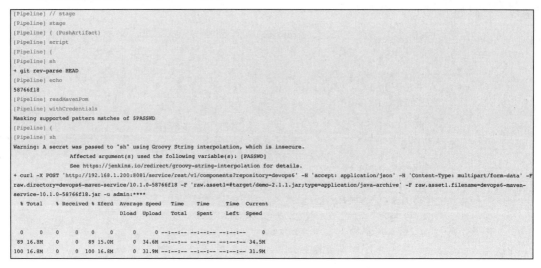

图 8-13　上传制品阶段日志

制品目录结构如图 8-14 所示。

图 8-14　制品目录结构

8.3　持续部署实践

8.3.1　准备工作

1．创建 Pipeline

创建一个持续部署流水线作业。这里笔者将一些通用的部署参数定义在 Jenkins 作业页面中展示，以便于参数的可视化。持续部署流水线参数如图 8-15 所示。

第 8 章 云主机环境持续部署

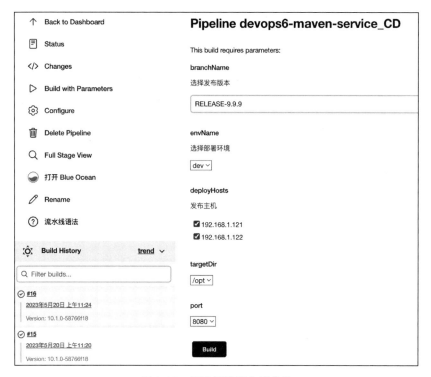

图 8-15 持续部署流水线参数

参数解释如下。

- ☑ branchName：版本分支名称，采用版本分支发布模式。
- ☑ envName：目标发布环境，以选项参数的方式选择。
- ☑ deployHosts：目标发布主机，以复选框的方式选择。
- ☑ targetDir：目标主机上的应用发布目录。
- ☑ port：服务在目标主机上的监听端口。

2．配置 Jenkinsfile

选择加载共享库中的 Jenkinsfile。笔者将 Jenkinsfile 和共享库代码放在一起，持续部署流水线用的 Jenkinsfile 名称是 cd.jenkinsfile。设置 SCM 为 Git 类型，填写文件保存的远程 Git 仓库地址、认证凭据、分支名称、脚本路径等，如图 8-16 所示。

3．加载共享库

笔者在共享库中编写了一些关于持续部署的模块，Jenkinsfile 中需要加载共享库并导入共享库中的模块。代码如下。

```
//加载共享库
@Library("devops06@main") _

//导入模块
def mygit = new org.devops.Gitlab()
def mydeploy = new org.devops.Deploy()
```

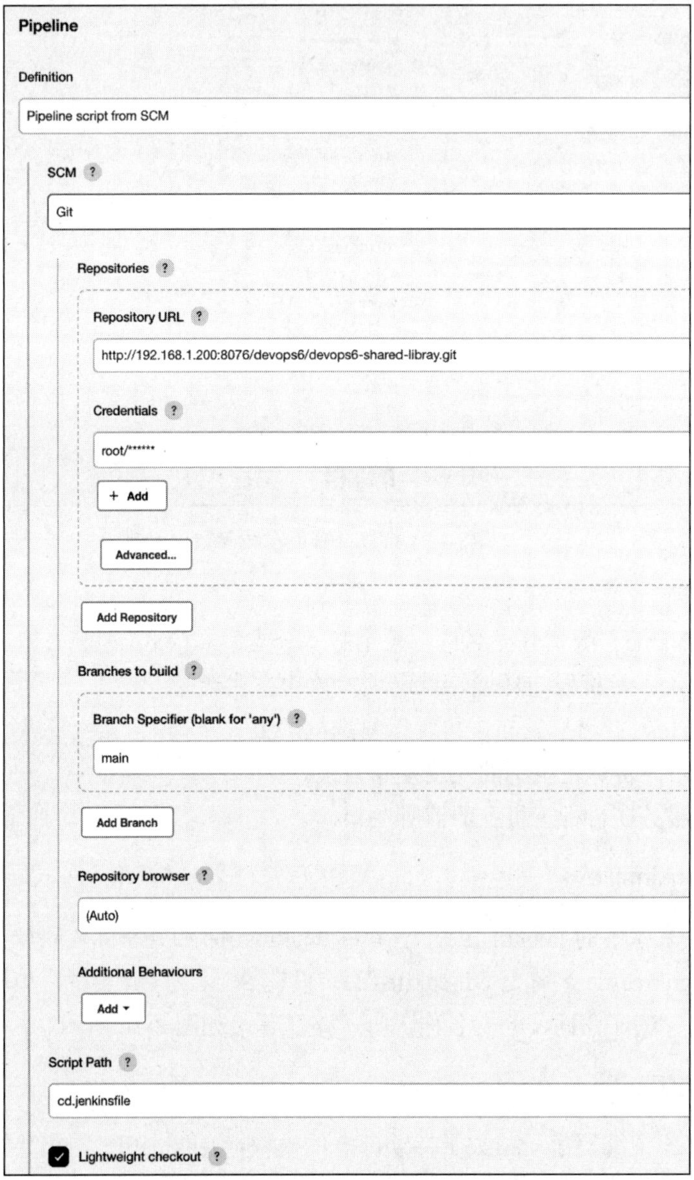

Pipeline

Definition

Pipeline script from SCM

SCM

Git

Repositories

Repository URL

http://192.168.1.200:8076/devops6/devops6-shared-libray.git

Credentials

root/******

+ Add

Advanced...

Add Repository

Branches to build

Branch Specifier (blank for 'any')

main

Add Branch

Repository browser

(Auto)

Additional Behaviours

Add ▼

Script Path

cd.jenkinsfile

☑ Lightweight checkout

图 8-16　Jenkinsfile 脚本配置

8.3.2 设置 Pipeline

设置 Pipeline 运行的构建节点、运行时选项、持续部署的多个阶段。持续部署阶段分为 GetArtifact（下载制品）和 Deploy（应用发布）阶段。代码如下。

```
//Pipeline
pipeline{
    agent { label "build"}          //在build标签节点运行
    options {
        skipDefaultCheckout true    //跳过默认代码下载
    }
    stages{
        stage("GetArtifact"){       //下载制品
            steps{
                script{
                    println("GetArtifact")
                }
            }
        }
        stage("Deploy"){            //应用发布
            steps{
                script{
                    println("Deploy")
                }
            }
        }
    }
}
```

1. 下载制品

在 Pipeline 的 GetArtifact 阶段下载制品，根据 Pipeline 输入的版本分支名称和提交 ID 等参数生成制品的下载链接，通过 wget 命令下载对应版本的制品。代码如下。

```
//check out
stage("GetArtifact"){
    steps{
        script{
            env.projectName = "${JOB_NAME}".split('_')[0] //devops6-maven-service
            env.groupName = "${env.projectName}".split('-')[0]  //devops6
```

```
            projectID = mygit.GetProjectIDByName(env.projectName,
env.groupName)
            commitID = mygit.GetBranchCommitID(projectID,
"${env.branchName}")
            println(commitID)
            appVersion = "${env.branchName}".split("-")[-1]   //9.9.9
            println(appVersion)
            currentBuild.description = "Version: ${appVersion}-${commitID}"

            repoUrl = "http://192.168.1.200:8081/repository/
${env.groupName}"   //nexus3
            env.artifactName = "${projectName}-${appVersion}-
${commitID}.jar"
            artifactUrl = "${repoUrl}/${env.projectName}/${appVersion}-
${commitID}/${env.artifactName}"
            sh "wget --no-verbose ${artifactUrl} && ls -l"

            env.releaseVersion = "${appVersion}-${commitID}"
        }
    }
}
```

在共享库 org/devops/Gitlab.groovy 中编写用于获取分支的提交 ID 功能的 3 个函数。代码如下。

```
//发起 HTTP 请求
def HttpReq(method, apiUrl){
    withCredentials([string(credentialsId:
'79f07070-bca3-4294-bcb4-c066449d6ffc',
                      variable: 'gitlabtoken')]) {
        response = sh returnStdout: true,
        script: """
            curl --location --request ${method} \
            http://192.168.1.200:8076/api/v4/${apiUrl} \
            --header "PRIVATE-TOKEN: ${gitlabtoken}"
        """
    }
    response = readJSON text: response - "\n"
    return response
}

//获取项目 ID
def GetProjectIDByName(projectName, groupName){
    apiUrl = "projects?search=${projectName}"
    response = HttpReq("GET", apiUrl)
```

```
    if (response != []){
        for (p in response) {
            if (p["namespace"]["name"] == groupName){
                return response[0]["id"]
            }
        }
    }
}

//获取分支提交 ID
def GetBranchCommitID(projectID, branchName){
    apiUrl = "projects/${projectID}/repository/branches/${branchName}"
    response = HttpReq("GET", apiUrl)
    return response.commit.short_id
}
```

HttpReq()函数是对 GitLab API 请求的封装，便于操作接口。GetProjectIDByName()函数根据 GitLab 仓库名称和组名称来查找对应的项目 ID，GetBranchCommitID()函数通过项目 ID 和分支名称参数来获取分支提交 ID。

下载制品阶段日志如图 8-17 所示。

```
[Pipeline] sh
Warning: A secret was passed to "sh" using Groovy String interpolation, which is insecure.
         Affected argument(s) used the following variable(s): [gitlabtoken]
         See https://jenkins.io/redirect/groovy-string-interpolation for details.
+ curl --location --request GET http://192.168.1.200:8076/api/v4/projects/5/repository/branches/RELEASE-10.1.0 --header 'PRIVATE-TOKEN: ****'
  % Total    % Received % Xferd  Average Speed   Time    Time     Time  Current
                                 Dload  Upload   Total   Spent    Left  Speed

  0     0    0     0    0     0      0      0 --:--:-- --:--:-- --:--:--     0
100  1072  100  1072    0     0  28210      0 --:--:-- --:--:-- --:--:-- 28210
[Pipeline] }
[Pipeline] // withCredentials
[Pipeline] readJSON
[Pipeline] echo
58766f18
[Pipeline] echo
10.1.0
[Pipeline] sh
+ wget --no-verbose http://192.168.1.200:8081/repository/devops6/devops6-maven-service/10.1.0-58766f18/devops6-maven-service-10.1.0-58766f18.jar
2023-05-20 11:24:50 URL:http://192.168.1.200:8081/repository/devops6/devops6-maven-service/10.1.0-58766f18/devops6-maven-service-10.1.0-58766f18.jar
[17669144/17669144] -> "devops6-maven-service-10.1.0-58766f18.jar.3" [1]
+ ls -l
总用量 155312
-rw-r--r-- 1 root root 17669144 5月  20 11:07 devops6-maven-service-10.1.0-58766f18.jar
```

图 8-17 在 Pipeline 中下载制品阶段日志

2．应用发布

编写一个 Shell 脚本用于服务控制，并将脚本存储在共享库的 resources/scripts/目录下。持续部署过程中会把这个脚本和应用一同发布到目标机器中。代码如下。

```bash
#!/bin/bash

#sh service.sh start
APPNAME=NULL
VERSION=NULL
PORT=NULL

#启动服务
start(){
   port_result=`netstat -anlpt | grep "${PORT}" || echo false`

   if [[ $port_result == "false" ]];then
      nohup java -jar -Dserver.port=${PORT} ${APPNAME}-${VERSION}.jar > ${APPNAME}.log.txt 2>&1 &
   else
     stop
     sleep 5
     nohup java -jar -Dserver.port=${PORT} ${APPNAME}-${VERSION}.jar > ${APPNAME}.log.txt 2>&1 &
   fi
}

#停止服务
stop(){
   pid=`netstat -anlpt | grep "${PORT}" | awk '{print $NF}' | awk -F '/' '{print $1}' | head -1`
   kill -15 $pid
}

#服务检查
check(){
   proc_result=`ps aux | grep java | grep "${APPNAME}" | grep -v grep || echo false`
   port_result=`netstat -anlpt | grep "${PORT}" || echo false`
   url_result=`curl -s http://localhost:${PORT} || echo false `

   if [[ $proc_result == "false" || $port_result == "false" || $url_result == "false" ]];then
       echo "server not running"
   else
       echo "ok"
   fi
}
```

```
#服务控制
case $1 in
   start)
       start
       sleep 5
       check
       ;;

   stop)
       stop
       sleep 5
       check
       ;;
   restart)
       stop
       sleep 5
       start
       sleep 5
       check
       ;;
   check)
       check
       ;;
   *)
       echo "sh service.sh {start|stop|restart|check}"
       ;;
esac
```

脚本中定义了 3 个变量，在发布过程中会替换对应的值。

- ☑ APPNAME：应用名称。
- ☑ VERSION：应用版本号。
- ☑ PORT：应用启动时监听的端口号。

在共享库 org/devops/Deploy.groovy 中编写 AnsibleDeploy()函数。该函数封装了使用 Ansible 进行应用发布的逻辑。首先，检查主机的连通性，然后清空目标机器的应用目录。使用 Ansible 将应用和启动脚本发布到目标机器的应用目录中，通过脚本启动应用程序。代码如下。

```
//发布制品
def AnsibleDeploy(){
   //将主机写入清单文件
   sh "rm -fr hosts "
```

```
        for (host in "${env.deployHosts}".split(',')){
            sh " echo ${host} >> hosts"
        }
        //ansible 发布jar
        sh """
            #主机连通性检测
            ansible "${env.deployHosts}" -m ping -i hosts

            #清理和创建发布目录
            ansible "${env.deployHosts}" -m shell -a "rm -fr ${env.targetDir}/${env.projectName}/* &&  mkdir -p ${env.targetDir}/${env.projectName} || echo file is exists"
            #发布应用包
            ansible "${env.deployHosts}" -m copy -a "src=${env.artifactName} dest=${env.targetDir}/${env.projectName}/${env.artifactName}"
        """
        //发布脚本
        fileData = libraryResource 'scripts/service.sh'
        //println(fileData)
        writeFile file: 'service.sh', text: fileData
        //sh "ls -a ; cat service.sh "

        sh """
            #修改变量
            sed -i 's#APPNAME=NULL#APPNAME=${env.projectName}#g' service.sh
            sed -i 's#VERSION=NULL#VERSION=${env.releaseVersion}#g' service.sh
            sed -i 's#PORT=NULL#PORT=${env.port}#g' service.sh

            #复制脚本
            ansible "${env.deployHosts}" -m copy -a "src=service.sh  dest=${env.targetDir}/${env.projectName}/service.sh"
            #启动服务
            ansible "${env.deployHosts}" -m shell -a "cd ${env.targetDir}/${env.projectName} ;source /etc/profile && sh service.sh start" -u root

            #检查服务
            sleep 10
            ansible "${env.deployHosts}" -m shell -a "cd ${env.targetDir}/${env.projectName} ;source /etc/profile && sh service.sh  check" -u root
        """
}
```

在Pipeline的Deploy阶段可以调用Deploy.groovy文件中的AnsibleDeploy()函数。代码如下。

```
stage("Deploy"){
   steps{
      script{
         mydeploy.AnsibleDeploy()
      }
   }
}
```

应用发布阶段日志如图 8-18 所示。

```
192.168.1.122 | CHANGED => {
    "ansible_facts": {
        "discovered_interpreter_python": "/usr/libexec/platform-python"
    },
    "changed": true,
    "checksum": "97c1a0ca48560a3ced9aa510615beb6154225505",
    "dest": "/opt/devops6-maven-service/service.sh",
    "gid": 0,
    "group": "root",
    "md5sum": "fb503617bb6407880a6bb60a15411e4f",
    "mode": "0644",
    "owner": "root",
    "secontext": "system_u:object_r:usr_t:s0",
    "size": 1369,
    "src": "/root/.ansible/tmp/ansible-tmp-1684553100.163275-85528109598488/source",
    "state": "file",
    "uid": 0
}
+ ansible 192.168.1.121,192.168.1.122 -m shell -a 'cd /opt/devops6-maven-service ;source /etc/profile && sh service.sh start' -u root
192.168.1.121 | CHANGED | rc=0 >>
ok
192.168.1.122 | CHANGED | rc=0 >>
ok
+ sleep 10
+ ansible 192.168.1.121,192.168.1.122 -m shell -a 'cd /opt/devops6-maven-service ;source /etc/profile && sh service.sh check' -u root
192.168.1.121 | CHANGED | rc=0 >>
ok
192.168.1.122 | CHANGED | rc=0 >>
ok
```

图 8-18 应用发布阶段日志

应用发布后，如果想将应用版本回滚怎么办？其实，如果制品库已经存储了上一个版本的应用包，可以填写上一个版本的分支，然后重新运行持续部署流水线发布上一个版本的包，这样就完成了应用版本的回滚。

8.4 本章小结

本章我们主要实践了企业中的私有云或者公有云环境下的持续集成和持续部署流水线。这里持续集成流水线是对持续集成部分章节知识的实践总结，持续部署部分通过

Ansible 完成了应用的发布。持续集成和持续部署流程图如图 8-19 所示。

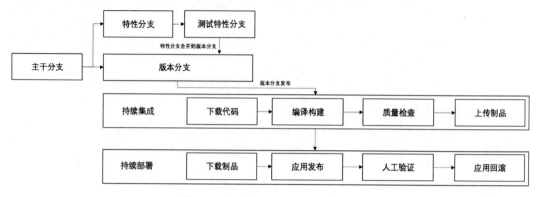

图 8-19　持续集成和持续部署流程图

第 9 章
Kubernetes 基础

Kubernetes 已经成为容器编排领域的事实标准,许多公司和组织都在使用 Kubernetes 管理容器化应用程序。本章笔者将聚焦于持续集成和持续部署相关的 Kubernetes 内容。

本章内容速览。
- ☑ Docker 容器基础
- ☑ Kubernetes 基础
- ☑ Kubernetes 部署策略
- ☑ 本章小结

9.1　Docker 容器基础

在传统的发布环境中存在多套环境,由于不同环境中的系统、应用配置层面的配置不同,导致应用程序的移植性很差。Docker 的核心特性是保证环境的一致性,将应用程序和运行环境打包成轻量的镜像,实现可移植性。

9.1.1　Docker 简介

Docker 是一个开源的容器化平台,它允许开发人员将应用程序和所有其依赖项打包成轻量级、可移植的镜像以便在任何地方运行。Docker 使用客户端/服务器(Client/Server)架构模式,Docker 架构组件图如图 9-1 所示。

Docker 相关术语如下。
- ☑ Docker Daemon:守护进程会处理复杂繁重的任务,如建立、运行、发布 Docker 容器。
- ☑ Docker Client:是一个二进制程序,主要用于 Docker 交互。它接收用户指令并且与 Docker 守护进程通信。

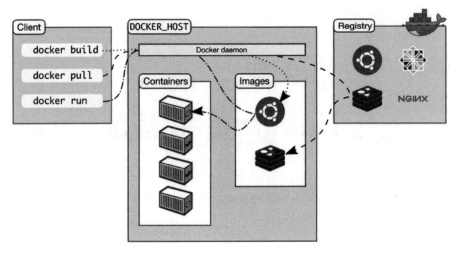

图 9-1 Docker 架构组件图

- ☑ Docker Images：包含应用程序和运行依赖项，每一个镜像由一系列的层组成。
- ☑ Docker Registry：仓库用来保存 Docker 镜像。
- ☑ Docker Container：Container 即容器，是 Docker 镜像的实例化。

Docker 的安装支持全平台，这里演示基于 Linux 安装 Docker 社区版本。安装命令如下：

```
#添加源
dnf config-manager --add-repo=https://download.docker.com/linux/centos/docker-ce.repo

#安装docker-ce
dnf list docker-ce

#启动服务
systemctl start docker
systemctl enable docker

#验证版本
docker -v
```

9.1.2 Docker 镜像构建

Docker 根据 Dockerfile 来构建生成 Docker 镜像。Dockerfile 是用于构建和部署 Docker 容器的文本文件，其中包含了指定容器如何构建和运行的命令。本节将简述 Dockerfile 语法和前后端项目的 Dockerfile 模板。

1. Dockerfile 语法

Dockerfile 中的关键字和语法具有一定的规则和约定，以便开发人员可以更轻松地编写和维护 Dockerfile。核心语法如下。

```
#FROM 指定基础镜像
FROM nginx:1.17.0

#MAINTAINER 维护者信息
MAINTAINER jenkins <jenkins@example.com>

#WORKDIR 指定工作目录
WORKDIR /dist

#ENV 构建时环境变量
ENV BUILD_TYPE=web

#ARG 构建镜像时可以通过 -build-arg 指定参数
ARG image_name=centos

#RUN 运行 Shell 命令
RUN echo hello

#ADD 添加文件
ADD run.sh /run.sh

#COPY 复制文件或目录
COPY index.html /usr/share/nginx/html

#VOLUME 定义共享存储卷或者临时存储卷
VOLUME /shareddata

#EXPOSE 指定容器开放端口
EXPOSE 80

#USER 指定容器运行时用户身份
USER root

#ENTRYPOINT 容器启动执行的命令
ENTRYPOINT ["/start.sh"]

#CMD 容器启动执行的命令
CMD ["yum", "-y", "install","wget"]
```

CMD 和 ENTRYPOINT 指令在工作方式上是有区别的。它们适用于不同的应用程序、环境和场景，都指定了在容器开始运行时执行的命令或者程序，但是当在 docker run 命令中声明参数时，CMD 指令将被覆盖，而 ENTRYPOINT 指令不会被覆盖，会将这些参数作为命令行附加参数。

2. 后端项目镜像构建

以 Java 项目为例。因为 Java 项目运行需要依赖 JDK 环境，首先导入基础镜像，然后将应用程序的 Jar 包复制到镜像中，最后通过启动命令运行。Dockerfile 语法如下。

```
#JDK1.8 版本镜像
FROM openjdk:8-jdk

#将应用包添加到容器
ADD app.jar /app.jar

#启动服务
ENTRYPOINT ["java", "-jar", "/app.jar"]
```

3. 前端项目镜像构建

前端项目发布的是网站静态资源文件，通常使用 Nginx 作为网站服务器。Dockerfile 语法如下。

```
#Nginx 镜像
FROM nginx:1.17.0

#复制网站资源到站点目录
COPY index.html /usr/share/nginx/html/
```

9.1.3　Docker 镜像管理

基于 Dockerfile 在本地生成 Docker 镜像后，可以通过命令上传到镜像仓库，以便于分发和共享。本节将简述 Harbor 镜像仓库的部署以及上传镜像用到的命令。

1. Docker 镜像仓库

Harbor 是一个用于存储和分发 Docker 镜像的企业级 Registry 服务器，可以用来构建企业内部的 Docker 镜像仓库。它具有企业需要的一些功能特性，如镜像同步复制、漏洞扫描和权限管理等。Harbor 的官网链接为 https://goharbor.io/。

Harbor 的部署需要使用 docker-compose。安装命令如下。

```
#下载docker-compose
curl -L "https://github.com/docker/compose/releases/download/1.26.2/
docker-compose-$(uname -s)-$(uname -m)" -o /usr/local/bin/docker-compose
#添加到系统环境
cp docker-compose-Linux-x86_64  /usr/local/bin/docker-compose
#添加可执行权限
chmod +x /usr/local/bin/docker-compose
#运行命令测试
docker-compose -v
Docker Compose version v2.17.2
```

Harbor 的部署包下载网址为 https://github.com/goharbor/harbor/releases。安装包分为 online（在线）和 offline（离线）两种类型。这里选择使用 online 的方式安装，读者可以按照实际情况自定义端口号，避免端口冲突。安装命令如下。

```
#解压安装包
tar zxf harbor-online-installer-v2.6.2.tgz
cd harbor
#编辑配置文件
vim harbor.yml
---
#修改主机名
hostname: 192.168.1.200

#端口配置
http:
  #port for http, default is 80. If https enabled, this port will redirect
to https port
  port: 8088
```

命令的运行结果如下。

```
----
#运行脚本进行安装
sh install.sh --with-chartmuseum  //开启 Helm Chart 仓库

[Step 4]: starting Harbor ...
Creating network "harbor_harbor" with the default driver
Creating harbor-log ... done
Creating registry         ... done
Creating harbor-portal ... done
Creating registryctl    ... done
Creating redis            ... done
Creating harbor-db      ... done
Creating harbor-core    ... done
```

```
Creating nginx            ... done
Creating harbor-jobservice ... done
✓ ----Harbor has been installed and started successfully.----
```

2. Docker 镜像命令

首先通过 docker build 命令基于 Dockerfile 构建生成镜像，然后通过 docker push 命令上传到镜像仓库。常用的镜像操作命令如下。

```
#登录镜像仓库
docker login 192.168.1.200:8088
Username: admin
Password:
WARNING! Your password will be stored unencrypted in /root/.docker/
config.json.
Configure a credential helper to remove this warning. See
https://docs.docker.com/engine/reference/commandline/login/#credentials
-store

Login Succeeded

#构建镜像
docker build -t 192.168.1.200:8088/devops6/devops6-maven-service:1.1.1 .

#上传镜像
docker push 192.168.1.200:8088/devops6/devops6-maven-service:1.1.1

#下载镜像
docker pull 192.168.1.200:8088/devops6/devops6-maven-service:1.1.1

#删除镜像
docker rmi 192.168.1.200:8088/devops6/devops6-maven-service:1.1.1

#注销登录
docker logout 192.168.1.200:8088
```

9.2 Kubernetes 基础

Kubernetes（K8s）是一个开源容器编排平台，可以扩展、自动部署和管理容器化应用程序。Kubernetes 编排应用程序的容器并实现高效管理和跟踪。

9.2.1 资源对象

Kubernetes 中存在的资源对象有很多，以常用的资源对象 Deployment、Service、Ingress 为例。Kubernetes 资源如图 9-2 所示。

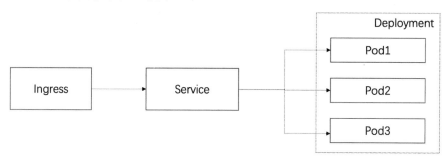

图 9-2　Kubernetes 资源

1. Deployment

Deployment 是经常使用的 Kubernetes 资源对象，通常用于描述 Pod。下面是一个使用 Deployment 的示例。其中创建了一个 ReplicaSet，负责启动 3 个 Nginx Pod 副本。

```
apiVersion: apps/v1
kind: Deployment
metadata:
  name: devops6-npm-service        #Deployment 名称
spec:
  replicas: 3                      #Pod 副本数量
  revisionHistoryLimit: 3          #历史版本保留数量
  selector:
    matchLabels:
      app: devops6-npm-service
  template:
    metadata:
      labels:
        app: devops6-npm-service   #Pod 标签
    spec:
      containers:
      - image: nginx:1.17.7        #镜像
        name: devops6-npm-service  #容器名称
        ports:
        - containerPort: 80        #端口
```

2. Service

Kubernetes Service 对象为一组 Pod 提供服务名称和 IP 地址。Service 提供 Pod 之间的发现和路由。Service 使用标签和选择器将 Pod 与其他应用程序相匹配。

```
apiVersion: v1
kind: Service
metadata:
  name: devops6-npm-service          #Service 名称
spec:
  type: ClusterIP                    #Service 类型
  selector:
    app: devops6-npm-service         #标签选择器
  ports:
  - name: http
    protocol: TCP
    port: 80
    targetPort: 80
```

Service 的类型如下。

- ☑ ClusterIP：默认的 Service 类型，用于在集群内部的 IP 地址上公开服务。只允许从集群内部访问。
- ☑ NodePort：通过每个 worker 节点上的静态绑定端口访问服务。
- ☑ LoadBalancer：通过云供应商的负载均衡器提供服务访问。

3. Ingress

Kubernetes Ingress 是一个对象，它定义了来自集群外部的流量（通常通过 HTTP 和 HTTPS 协议）路由给内部 Kubernetes 集群中的 Service。

```
apiVersion: networking.k8s.io/v1
kind: Ingress
metadata:
  name: devops6-npm-service                    #Ingress 名称
  annotations:
    kubernetes.io/ingress.class: nginx         #nginx 控制器
spec:
  rules:
  - host: devops.test.com                      #域名
    http:
      paths:
      - path: /
        pathType: Prefix
        backend:
```

```yaml
  service:
    name: devops6-npm-service         #Service 名称
    port:
      name: http
```

9.2.2 Kubectl 工具发布

Kubectl 是特定于 Kubernetes 的命令行工具，可以与 API 通信并管理 Kubernetes 集群。例如，在 Kubernetes 平台上创建、管理、删除资源。下面将演示使用 Kubectl 工具发布应用和回滚应用。

1．发布应用

当编写完资源清单文件后，可以通过 kubectl create 或 kubectl apply 命令进行发布。命令如下。

```
#创建资源
kubectl create -f deployment.yaml

#应用资源
kubectl apply -f deployment.yaml
```

2．回滚应用

当发布的应用不符合预期时，可以通过 kubectl rollout 命令进行回滚。命令如下。

```
#查看历史
kubectl rollout history deployment/nginx

#查看具体某一个历史版本信息
kubectl rollout history deployment/nginx --revision=2

#回滚上个版本
kubectl rollout undo deployment/nginx

#回滚指定版本
kubectl rollout undo deployment/nginx --to-revision=2
```

9.2.3 Helm 工具发布

Helm 是一种 Kubernetes 工具，通过将清单文件合并到单个可重用包中，自动创建、打包、配置和部署 Kubernetes 应用程序。

Helm Chart 是一个包，其中包含将应用程序部署到 Kubernetes 集群所需的所有资源，包括用于定义应用程序所需状态的 Deployment、Service、Secret 和 ConfigMap 等 YAML 配置文件。Helm Chart 将 YAML 文件和模板打包在一起，这些文件和模板可用于根据参数化值 values.yaml 生成配置文件。支持通过不同的 values 配置文件适配不同的环境。

1. 发布应用

Helm Chart 的源码通常保存在 Git 仓库中，可以进行打包将 Chart 上传到仓库，然后安装或更新到环境。命令如下。

```
#打包 Helm Chart
helm package ${appName} --version ${chartVersion}
#上传 Chart 到仓库
helm cm-push ${appName}-${chartVersion}.tgz devops6repo
#添加 Helm repo
helm repo add devops6repo http://192.168.1.200:8088/chartrepo/devops6/
#更新 Helm repo
helm repo update devops6repo
#下载 Chart
helm pull devops6repo/${env.appName} --version ${env.chartVersion}
#更新或安装 Chart
helm upgrade --install --create-namespace RELEASE_NAME chart.tgz
```

2. 回滚应用

通过 helm rollback 命令指定版本可以回滚应用。命令如下。

```
#查看历史
helm history [RELEASE]
#回滚
helm rollback [RELEASE] [REVISION]
```

9.3　Kubernetes 部署策略

之前，开发人员在部署更改和更新时会使应用程序出现问题或导致停机。蓝绿部署方法和金丝雀部署方法可以在部署应用程序时避免给用户带来任何停机或中断问题。

9.3.1　滚动更新

Kubernetes 滚动更新的基本原理是使用 rolling-update 策略，该策略允许在集群中逐步运行新版本的应用 Pod，而不必一次性将其全部部署到集群中。在 Deployment 中配置滚动更新如下。

```
#Deployment
apiVersion: apps/v1beta2
kind: Deployment
metadata:
  name: nginx
  labels:
    app: nginx
spec:
  replicas: 1
  selector:
    matchLabels:
      app: nginx
  strategy:
    type: RollingUpdate          #滚动更新
    rollingUpdate:
      maxUnavailable: 1          #最多 1 个 Pod 处于不可工作状态
      maxSurge: 2                #升级时可以比预期多出两个 Pod
```

参数解释如下。

- maxUnavailable：指定更新时最大不可用的 Pod 数量或者百分比，默认值为 25%。
- maxSurge：指定更新时可以多出预期最大的 Pod 数量，默认值为 25%。

9.3.2 零停机部署

蓝绿部署是减少部署停机时间的方法之一，存在蓝和绿两个环境，升级过程通过负载均衡器进行控制。当升级过程中出现问题时可以快速切换到旧版本。

1．初始发布

初始发布中，所有用户流量访问都在绿环境的 version 1.0 中，如图 9-3 所示。

图 9-3　初始发布流量图

2．新版本发布

version 2.0 发布后，流量依然在 version 1.0。此时开发人员可以在 version 2.0 中进行功能测试和验证，如图 9-4 所示。

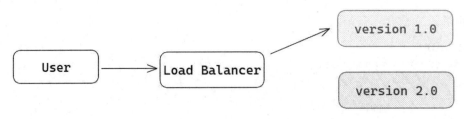

图 9-4　新版本发布流量图

3. 流量切换

当开发人员完成新功能的验证后，操作 Load Balancer 将流量切换到 version 2.0。此时所有用户可以访问新版本。旧版本不会立即删除，直到新版本稳定运行后再计划删除，如图 9-5 所示。

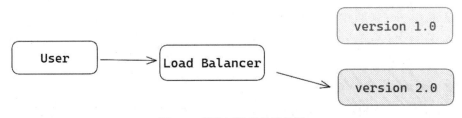

图 9-5　流量切换后的流量图

4. 终态完成

流量切换完成后，所有用户访问新版本——version 2.0，如图 9-6 所示。

图 9-6　终态流量图

金丝雀部署方法与蓝绿部署方法有些类似，但实现的方法略有不同。在金丝雀部署中，首先只更新一小部分服务器或节点，然后再完成其他节点更新。

9.4　本章小结

本章我们从持续集成持续部署的角度讲解了 Docker 和 Kubernetes 的基础知识。这便于读者对后续章节持续部署管道中运行步骤的理解。Kubernetes 的内容还有很多，本节笔者仅聚焦于持续集成和持续部署用到的知识。

第 10 章
Kubernetes 持续部署

Kubernetes 是一个开源容器编排平台，可以自动部署、管理和扩展容器化应用程序。换句话说，Kubernetes 是一个集群管理系统，可以保持工作负载运行，这在 DevOps 中很重要。随着云原生使用需求的爆炸式增长，Kubernetes 已发展成增长最快的基础设施平台。

本章内容速览。

- ☑ 持续集成流水线
- ☑ 基于 Kubectl 持续部署
- ☑ 基于 Helm 持续部署
- ☑ 本章小结

10.1 持续集成流水线

基于 Kubernetes 环境实践持续集成流水线，该流水线包含下载代码、编译构建、质量检查、镜像构建阶段。基于 Kubernetes 的持续集成流水线和传统云主机的流水线的主要区别是生成 Docker 镜像并上传到镜像仓库，持续集成流水线阶段如图 10-1 所示。

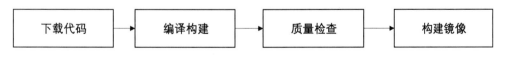

图 10-1 持续集成流水线阶段

10.1.1 准备工作

1. 创建 Pipeline

创建一个持续集成流水线作业。配置代码地址、分支名称、构建命令等参数，如图 10-2 所示。

```
┌─────────────────────────────────────────────────────────────────────┐
│  ↑  Back to Dashboard      Pipeline devops6-npm-service_K8SCI       │
│  ☰  Status                 This build requires parameters:          │
│  </> Changes               srcUrl                                   │
│  ▷  Build with Parameters  [http://192.168.1.200:8076/devops6/devops6-npm-service.git ∨] │
│  ⚙  Configure              buildShell                               │
│  🗑  Delete Pipeline        [npm install && npm run build         ] │
│  🔍 Full Stage View         branchName                              │
│     SonarQube              [RELEASE-1.1.1 ∨]                        │
│  ◉  打开 Blue Ocean         [ Build ]                                │
│  ✎  Rename                                                          │
└─────────────────────────────────────────────────────────────────────┘
```

图 10-2 持续集成流水线参数

参数解释如下。

- ☑ srcUrl：项目代码库地址。
- ☑ buildShell：项目构建命令。
- ☑ branchName：项目分支名称。

2．配置 Jenkinsfile

选择加载共享库中的 Jenkinsfile。笔者将 Jenkinsfile 和共享库代码放到一起，持续集成流水线用的 Jenkinsfile 名称是 k8sci.jenkinsfile。SCM 选项用于选择仓库类型，这里选择 Git 类型，填写文件保存的远程 Git 仓库地址、认证凭据、分支名称、脚本路径等，如图 10-3 所示。

3．加载共享库

笔者在共享库中编写了一些关于持续集成的模块，Jenkinsfile 中需要加载共享库并导入共享库中的模块。代码如下。

```
//加载共享库
@Library("devops06@main") _

//导入模块
def build = new org.devops.Build()
def sonar = new org.devops.Sonar()
```

```
def checkout = new org.devops.Checkout()
def mygit = new org.devops.Gitlab()
```

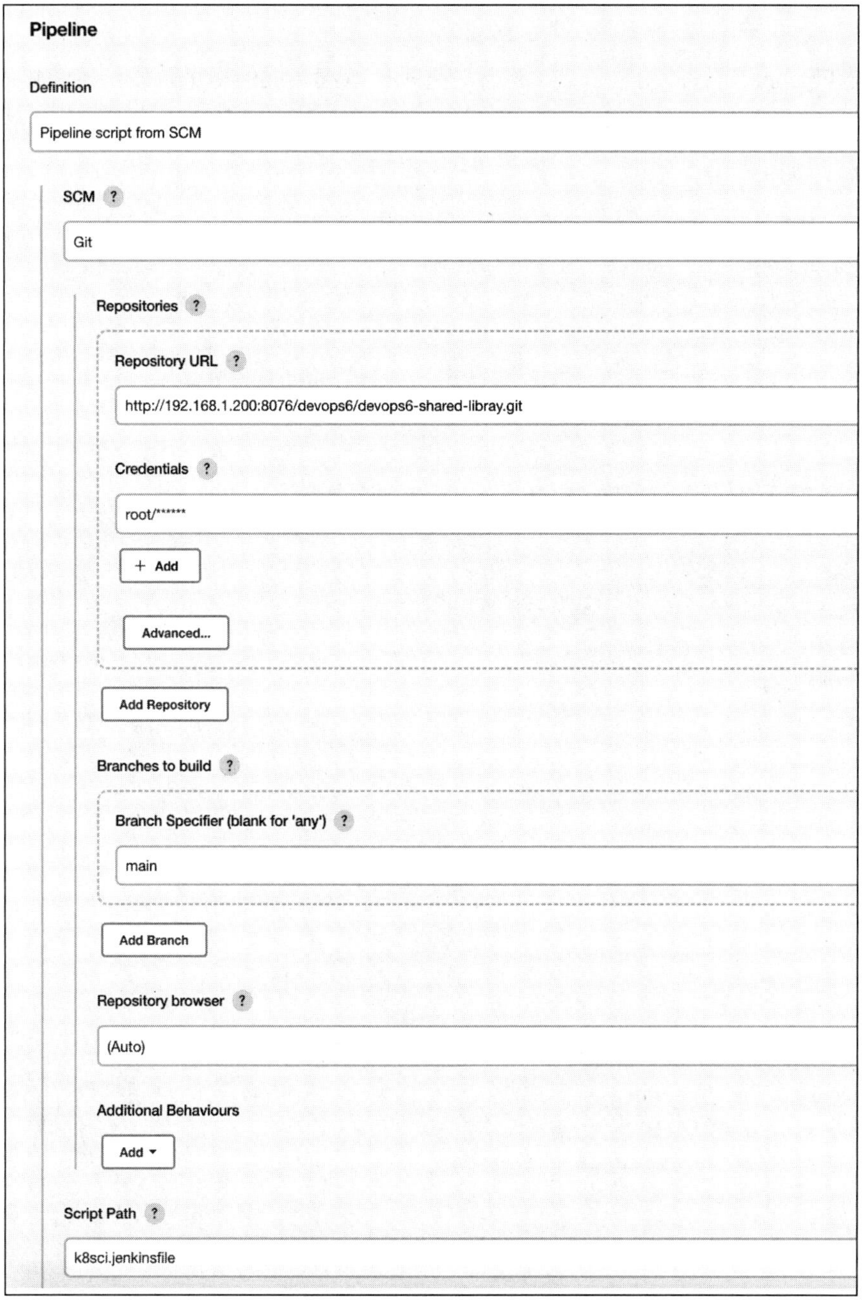

图 10-3　Jenkinsfile 脚本配置

10.1.2 设置 Pipeline

设置 Pipeline 运行的构建节点、运行时选项、持续集成的多个阶段。持续集成阶段包括 CheckOut（下载代码）、Build（编译构建）、CodeScan（质量检查）、ImageBuild（构建镜像）。代码如下。

```groovy
//持续集成流水线
pipeline {
    agent { label "build"}              //运行在build标签的节点
    options {
        skipDefaultCheckout true        //跳过默认的代码检出
    }
    stages{
        stage("CheckOut"){              //下载代码
            steps{
                script{
                    println("checkout")
                }
            }
        }
        stage("Build"){                 //编译构建
            steps{
                script{
                    println("build")
                }
            }
        }
        stage("CodeScan"){              //质量检查
            steps{
                script{
                    println("codescan")
                }
            }
        }
        stage("ImageBuild"){            //构建镜像
```

```
        steps{
            script{
                println("build docker image")
            }
        }
    }
}
```

1. 下载代码

在共享库 org/devops/Build.groovy 中编写 CheckOut()函数。该函数调用 Jenkins 内置语法 checkout 语句来下载项目的代码并运行 Shell 命令进行验证代码的下载。代码如下。

```
//check out
def CheckOut(){
   println("CheckOut")
   checkout([$class: 'GitSCM',
      branches: [[name: "${env.branchName}"]],
      extensions: [],
      userRemoteConfigs: [[credentialsId: 'gitlab-admin',
      url: "${params.srcUrl}"]]])
   sh "ls -l " //验证
}
```

在 Pipeline 的 CheckOut 阶段可以直接调用 Build.groovy 文件中定义的 CheckOut()函数。代码如下。

```
stage("CheckOut"){
   steps{
      script{
         build.CheckOut()
      }
   }
}
```

下载代码阶段的日志如图 10-4 所示。

```
[Pipeline] echo
CheckOut
[Pipeline] checkout
The recommended git tool is: NONE
using credential gitlab-admin
Fetching changes from the remote Git repository
 > git rev-parse --resolve-git-dir /opt/jenkinsagent/workspace/devops6-npm-service_K8SCI/.git # timeout=10
 > git config remote.origin.url http://192.168.1.200:8076/devops6/devops6-npm-service.git # timeout=10
Fetching upstream changes from http://192.168.1.200:8076/devops6/devops6-npm-service.git
 > git --version # timeout=10
 > git --version # 'git version 2.31.1'
using GIT_ASKPASS to set credentials
 > git fetch --tags --force --progress -- http://192.168.1.200:8076/devops6/devops6-npm-service.git +refs/heads/*:refs/remotes/origin/* #
Checking out Revision fbc1e7bbcef4c7ff9ba25a3ca4d19fbbb70cd943 (origin/RELEASE-2.1.1)
Commit message: "Update index.html"
 > git rev-parse origin/RELEASE-2.1.1^{commit} # timeout=10
 > git config core.sparsecheckout # timeout=10
 > git checkout -f fbc1e7bbcef4c7ff9ba25a3ca4d19fbbb70cd943 # timeout=10
 > git rev-list --no-walk fbc1e7bbcef4c7ff9ba25a3ca4d19fbbb70cd943 # timeout=10
[Pipeline] sh
+ ls -l
总用量 1100
drwxr-xr-x   2 root root     188 5月  27 10:13 build
drwxr-xr-x   2 root root      59 5月  27 10:13 config
-rw-r--r--   1 root root    1030 5月  27 10:47 Deployment.yaml
drwxr-xr-x   3 root root      38 5月  27 10:49 dist
-rw-r--r--   1 root root      58 5月  27 10:13 Dockerfile
-rw-r--r--   1 root root     304 5月  27 10:49 index.html
drwxr-xr-x 829 root root   24576 5月  27 10:49 node_modules
-rw-r--r--   1 root root    1740 5月  27 10:13 package.json
-rw-r--r--   1 root root 1057054 5月  27 10:50 package-lock.json
-rw-r--r--   1 root root     464 5月  27 10:13 README.md
-rw-r--r--   1 root root     431 5月  27 10:13 sonar-project.properties
drwxr-xr-x   4 root root      68 5月  27 10:13 src
drwxr-xr-x   2 root root      22 5月  27 10:13 static
```

图 10-4　下载代码阶段日志

2. 编译构建

在共享库 org/devops/Build.groovy 中编写 Build()函数。该函数调用 Shell 命令进行编译项目构建。例如，此处将会加载在 Jenkins 页面定义的 buildShell 参数的值，即运行 mvn clean package 命令进行项目编译构建。代码如下。

```
//编译构建
def Build(){
    println("Build")
    sh "${params.buildShell}"   //构建命令
}
```

在 Pipeline 的 Build 阶段可以直接调用 Build.groovy 文件中定义的 Build()函数。代码如下。

```
stage("Build"){
    steps{
        script{
            build.Build()
```

```
        }
    }
}
```

编译构建阶段的日志如图 10-5 所示。

```
(node:88897) Warning: Accessing non-existent property 'test' of module exports inside circular dependency
(node:88897) Warning: Accessing non-existent property 'to' of module exports inside circular dependency
(node:88897) Warning: Accessing non-existent property 'toEnd' of module exports inside circular dependency
(node:88897) Warning: Accessing non-existent property 'touch' of module exports inside circular dependency
(node:88897) Warning: Accessing non-existent property 'uniq' of module exports inside circular dependency
(node:88897) Warning: Accessing non-existent property 'which' of module exports inside circular dependency
Hash:   [1m194cad9c6f44f07fb00e [39m [22m
Version: webpack [1m3.12.0 [39m [22m
Time:   [1m4250 [39m [22mms
                                     [1mAsset [39m [22m        [1mSize [39m [22m  [1mChunks [39m [22m            [1m [39m [
Names [39m [22m
              [1m [32mstatic/js/app.2f2e5edd9af2c59aa514.js [39m [22m    11.5 kB       [1m0 [39m [22m  [1m [32m[emitted]
             [1m [32mstatic/js/vendor.6fb2cc639d10e850d042.js [39m [22m     102 kB       [1m1 [39m [22m  [1m [32m[emitted]
           [1m [32mstatic/js/manifest.2ae2e69a05c33dfc65f8.js [39m [22m    857 bytes      [1m2 [39m [22m  [1m [32m[emitted]
        [1m [32mstatic/css/app.30790115300ab27614ce176899523b62.css [39m [22m   432 bytes      [1m0 [39m [22m  [1m [32m[emitted]
    [1m [32mstatic/css/app.30790115300ab27614ce176899523b62.css.map [39m [22m   797 bytes      [1m0 [39m [22m  [1m [32m[emitted]
          [1m [32mstatic/js/app.2f2e5edd9af2c59aa514.js.map [39m [22m     21.7 kB       [1m0 [39m [22m  [1m [32m[emitted]
         [1m [32mstatic/js/vendor.6fb2cc639d10e850d042.js.map [39m [22m     598 kB       [1m1 [39m [22m  [1m [32m[emitted]
       [1m [32mstatic/js/manifest.2ae2e69a05c33dfc65f8.js.map [39m [22m     4.97 kB       [1m2 [39m [22m  [1m [32m[emitted]
                                     [1m [32mindex.html [39m [22m    532 bytes      [1m [39m [22m  [1m [32m[emitted]

Build complete.

Tip: built files are meant to be served over an HTTP server.
   Opening index.html over file:// won't work.
```

图 10-5　编译构建阶段日志

3．质量检查

在共享库 org/devops/Sonar.groovy 中编写 SonarScannerByPlugin()函数。该函数封装了 SonarQube 插件提供的 withSonarQubeEnv 语句进行代码扫描。由于代码扫描需要使用 SonarQube 用户认证，读者需要将其保存在 Jenkins 凭据中。代码如下：

```
//SonarScannerByPlugin
def SonarScannerByPlugin(){
    withSonarQubeEnv(credentialsId:
'fdf4362a-69e7-4014-8fa7-80b1ba268588') {
        withCredentials([[usernamePassword(credentialsId: '9ff42b72-597e-
49dd-a62f-2553d48304fc', passwordVariable: 'SONAR_PASSWD',usernameVariable:
'SONAR_USER')]) {
            sh """
                sonar-scanner \
                -Dsonar.login=${SONAR_USER} \
                -Dsonar.password=${SONAR_PASSWD} \
```

```
                -Dsonar.host.url=http://192.168.1.200:9000 \
                -Dsonar.branch.name=${env.branchName}
            """
        }
    }
}
```

在 Pipeline 的 CodeScan 阶段可以调用 Sonar.groovy 文件中的 SonarScannerByPlugin() 函数。代码如下。

```
stage("CodeScan"){
    steps{
        script{
            sonar.SonarScannerByPlugin()
        }
    }
}
```

质量检查阶段日志如图 10-6 所示。

```
INFO: Sensor IaC Docker Sensor [iac]
INFO: 0 source files to be analyzed
INFO: 0/0 source files have been analyzed
INFO: Sensor IaC Docker Sensor [iac] (done) | time=78ms
INFO: ------------- Run sensors on project
INFO: Sensor Analysis Warnings import [csharp]
INFO: Sensor Analysis Warnings import [csharp] (done) | time=1ms
INFO: Sensor Zero Coverage Sensor
INFO: Sensor Zero Coverage Sensor (done) | time=0ms
INFO: SCM Publisher SCM provider for this project is: git
INFO: SCM Publisher 3 source files to be analyzed
INFO: SCM Publisher 3/3 source files have been analyzed (done) | time=172ms
INFO: CPD Executor Calculating CPD for 0 files
INFO: CPD Executor CPD calculation finished (done) | time=0ms
INFO: Load New Code definition
INFO: Load New Code definition (done) | time=212ms
INFO: Analysis report generated in 292ms, dir size=126.3 kB
INFO: Analysis report compressed in 10ms, zip size=18.2 kB
INFO: Analysis report uploaded in 156ms
INFO: ANALYSIS SUCCESSFUL, you can find the results at: http://192.168.1.200:9000/dashboard?id=devops6-npm-service&br
INFO: Note that you will be able to access the updated dashboard once the server has processed the submitted analysis
INFO: More about the report processing at http://192.168.1.200:9000/api/ce/task?id=AYhbHfX1sVnW6oTXeNgW
INFO: Analysis total time: 10.621 s
INFO: ------------------------------------------------------------------------
INFO: EXECUTION SUCCESS
INFO: ------------------------------------------------------------------------
INFO: Total time: 12.018s
INFO: Final Memory: 19M/80M
INFO: ------------------------------------------------------------------------
```

图 10-6　质量检查阶段日志

4. 构建镜像

本节，笔者整理了关于 Docker 镜像操作的相关命令。持续集成流水线中的镜像构建阶段主要运行 docker 命令登录镜像仓库、构建镜像、上传镜像、删除镜像。代码如下：

```
stage("ImageBuild"){
    steps{
        script{

            //获取仓库信息
            appName = "${JOB_NAME}".split('_')[0] //devops6-npm-service
            repoName = appName.split('-')[0]   //devops6
            //获取提交 ID
            commitID = checkout.GetCommitID()
            println(commitID)
            //定义镜像名称
            imageName = "${repoName}/${appName}"
            imageTag = "${env.branchName}-${commitID}"
            env.fullImageName = "192.168.1.200:8088/${imageName}:${imageTag}"

            //构建镜像
sh """
            #登录镜像仓库
            docker login -u admin -p Harbor12345 192.168.1.200:8088

            #构建镜像
            docker build -t ${env.fullImageName} .

            #上传镜像
            docker push ${env.fullImageName}

            #删除镜像
            sleep 2
            docker rmi ${env.fullImageName}
            """
        }
    }
}
```

构建镜像阶段日志如图 10-7 所示。

```
[Pipeline] sh
+ docker login -u admin -p Harbor12345 192.168.1.200:8088
WARNING! Using --password via the CLI is insecure. Use --password-stdin.
WARNING! Your password will be stored unencrypted in /root/.docker/config.json.
Configure a credential helper to remove this warning. See
https://docs.docker.com/engine/reference/commandline/login/#credentials-store

Login Succeeded
+ docker build -t 192.168.1.200:8088/devops6/devops6-npm-service:RELEASE-2.1.1-fbc1e7bb .
#1 [internal] load build definition from Dockerfile
#1 transferring dockerfile: 31B done
#1 DONE 0.0s

#2 [internal] load .dockerignore
#2 transferring context: 2B done
#2 DONE 0.0s

#3 [internal] load metadata for docker.io/library/nginx:1.17.0
#3 DONE 15.3s

#4 [1/2] FROM docker.io/library/nginx:1.17.0@sha256:bdbf36b7f1f77ffe7bd2a32e59235dff6ecf131e3b6b5b96061c652f30685f3a
#4 DONE 0.0s

#5 [internal] load build context
#5 transferring context: 32B done
#5 DONE 0.0s

#6 [2/2] COPY index.html /usr/share/nginx/html/
#6 CACHED

#7 exporting to image
#7 exporting layers done
#7 writing image sha256:a661e79c85efcf6b9e6887a5ed94ffe6e39cd93c1f2447a7059b5381b2703453 done
#7 naming to 192.168.1.200:8088/devops6/devops6-npm-service:RELEASE-2.1.1-fbc1e7bb done
#7 DONE 0.0s
+ docker push 192.168.1.200:8088/devops6/devops6-npm-service:RELEASE-2.1.1-fbc1e7bb
The push refers to repository [192.168.1.200:8088/devops6/devops6-npm-service]
a1023414d440: Preparing
d7acf794921f: Preparing
d9569ca04881: Preparing
cf5b3c6798f7: Preparing
```

图 10-7 构建镜像阶段日志

注意：

笔者在运行的 Jenkins Agent 节点上安装了 Docker，所以可以直接运行 docker 命令，如果读者在运行流水线时没有找到 docker 命令，则需要在 Jenkins Agent 节点手动安装。

10.1.3 启用 GitOps

GitOps 是基于 Git 实现 CI/CD 的一种实践方式。笔者将 Kuberentes 部署清单文件托管到 Git 代码库中进行版本控制，这样既可以将环境配置变更历史记录下来，还可以通过 Git

第 10 章　Kubernetes 持续部署

仓库对变更代码进行预览和审查，并在必要时进行版本回滚。

通常情况下，Kuberente 部署清单文件中的镜像更新是手动完成的。在此，笔者将通过 GitLab API 在持续集成管道中对部署清单文件中的镜像进行更新。基于 GitLab API 笔者写了一些 Groovy 函数存放在共享库 org/devops/Gitlab.groovy 文件中。代码如下。

```groovy
package org.devops

//获取文件内容
def GetRepoFile(projectId,filePath, branchName ){
    //GET /projects/:id/repository/files/:file_path/raw
    apiUrl = "/projects/${projectId}/repository/files/${filePath}/raw?ref=${branchName}"
    response = HttpReq('GET', apiUrl)
    return response
}

//更新文件内容
def UpdateRepoFile(projectId,filePath,fileContent, branchName){
    apiUrl = "projects/${projectId}/repository/files/${filePath}"
    reqBody = """{"branch": "${branchName}","encoding":"base64","content": "${fileContent}", "commit_message": "update a new file"}"""
    response = HttpReqByPlugin('PUT',apiUrl,reqBody)
    println(response)

}

//创建文件
def CreateRepoFile(projectId,filePath,fileContent, branchName){
    apiUrl = "projects/${projectId}/repository/files/${filePath}"
    reqBody = """{"branch": "${branchName}","encoding":"base64","content": "${fileContent}", "commit_message": "update a new file"}"""
    response = HttpReqByPlugin('POST',apiUrl,reqBody)
    println(response)
}

//封装 HTTP
def HttpReqByPlugin(reqType, reqUrl,reqBody ){
    def gitServer = "http://192.168.1.200:8076/api/v4"
    withCredentials([string(credentialsId: '79f07070-bca3-4294-bcb4-c066449d6ffc', variable: 'GITLABTOKEN')]) {
        response = httpRequest acceptType: 'APPLICATION_JSON_UTF8',
                    consoleLogResponseBody: true,
                    contentType: 'APPLICATION_JSON_UTF8',
```

```
                customHeaders: [[maskValue: false,
                                 name: 'PRIVATE-TOKEN',
                                 value: "${GITLABTOKEN}"]],
                httpMode: "${reqType}",
                url: "${gitServer}/${reqUrl}",
                wrapAsMultipart: false,
                requestBody: "${reqBody}"
    }
    return response
}


//发起HTTP请求
def HttpReq(method, apiUrl){
    withCredentials([string(credentialsId: '79f07070-bca3-4294-bcb4-c066449d6ffc', variable: 'gitlabtoken')]) {
        response = sh returnStdout: true,
        script: """
            curl --location --request ${method} \
            http://192.168.1.200:8076/api/v4/${apiUrl} \
            --header "PRIVATE-TOKEN: ${gitlabtoken}"
        """
    }
    try {
        response = readJSON text: response - "\n"
    } catch(e){
        println(e)
    }
    return response
}

//获取项目ID
def GetProjectIDByName(projectName, groupName){
    apiUrl = "projects?search=${projectName}"
    response = HttpReq("GET", apiUrl)
    if (response != []){
        for (p in response) {
            if (p["namespace"]["name"] == groupName){
                return response[0]["id"]
            }
        }
    }
}
```

```
//获取分支提交 ID
def GetBranchCommitID(projectID, branchName){
    apiUrl = "projects/${projectID}/repository/branches/${branchName}"
    response = HttpReq("GET", apiUrl)
    return response.commit.short_id
}
```

注意:

笔者分别使用封装 curl 命令和 Jenkins 的 HTTP Request Plugin 请求 GitLab API。所以，读者在使用这段代码时，需要提前在 Jenkins 平台安装 HTTP Request Plugin。

下面笔者分别演示基于 Kubectl 工具和基于 Helm 工具的两种实现方式。

1. 基于 Kubectl 工具

创建一个 Git 仓库，用于存储 Kubernetes 部署清单文件。仓库内容如图 10-8 所示。

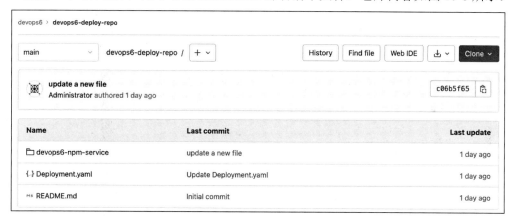

图 10-8　Kubernetes 部署清单仓库

在仓库的根目录创建一个 Deployment.yaml 文件，此文件包含应用部署用到的 Ingress、Service、Deployment 资源清单。笔者将此文件作为一个模板，内容如下。

```
apiVersion: apps/v1
kind: Deployment
metadata:
  name: devops6-npm-service
spec:
  replicas: 3
  revisionHistoryLimit: 3
  selector:
    matchLabels:
```

```
      app: devops6-npm-service
  template:
    metadata:
      labels:
        app: devops6-npm-service
    spec:
      containers:
      - image: __IMAGE_NAME__        //镜像名称的变量
        name: devops6-npm-service
        ports:
        - containerPort: 80
---
apiVersion: v1
kind: Service
metadata:
  name: devops6-npm-service
spec:
  type: ClusterIP
  selector:
    app: devops6-npm-service
  ports:
  - name: http
    protocol: TCP
    port: 80
    targetPort: 80
---
apiVersion: networking.k8s.io/v1
kind: Ingress
metadata:
  name: devops6-npm-service
  annotations:
    kubernetes.io/ingress.class: nginx
spec:
  rules:
  - host: devops.test.com
    http:
      paths:
      - path: /
        pathType: Prefix
        backend:
          service:
            name: devops6-npm-service
            port:
              name: http
```

在持续集成流水线中，首先下载 Deployment.yaml 模板文件，然后替换镜像变量名称，最后基于当前的版本生成新的部署文件。流水线代码如下。

```
//更新版本文件
stage("UpdateEnvFile"){
    steps{
        script {
            projectId = 9                    //存放清单文件的 Git 仓库的项目 ID
            fileName = "Deployment.yaml"     //模板文件名称
            branchName = "main"
            //下载模板文件
            fileData = mygit.GetRepoFile(projectId,fileName, branchName )
            sh "rm -fr ${fileName}"

            //模板文件内容保存到本地
            writeFile file: fileName , text: fileData
            env.deployFile = fileName

            //替换镜像
            sh "sed -i 's#__IMAGE_NAME__#${env.fullImageName}#g' ${env.deployFile} "
            sh "ls -l ; cat ${fileName}"

            //创建/更新发布文件
            newYaml = sh returnStdout: true, script: "cat ${env.deployFile}"
            println(newYaml)

            //更新 GitLab 文件内容
            base64Content = newYaml.bytes.encodeBase64().toString()
            appName = "${JOB_NAME}".split('_')[0]  //devops6-npm-service
            env.groupName = appName.split('-')[0]   //devops6
            env.projectName = appName

            //如果文件不存在则创建，存在则更新
            try {
                mygit.UpdateRepoFile(projectId,
                        "${env.projectName}%2f${env.branchName}.yaml",
                        base64Content,
                        "main")
            } catch(e){
                mygit.CreateRepoFile(projectId,
                        "${env.projectName}%2f${env.branchName}.yaml",
```

```
                    base64Content,
                    "main")
            }
        }
    }
}
```

将上面代码扩展到持续集成流水线后，如果运行成功，则可以进入 GitOps 仓库中找到对应的发布版本文件。查看预览版本文件的内容可以看到镜像已经被替换成功。后续持续部署流水线则可以根据版本号下载此部署文件，然后通过 kubectl 命令进行发布，如图 10-9 所示。

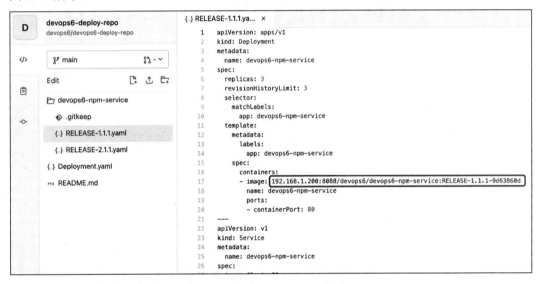

图 10-9　Kubernetes 部署清单仓库内容

2. 基于 Helm 工具

Helm 是 Kubernetes 的包管理工具。与 Kubectl 不同的是，Helm 需要通过 helm create 命令创建一个 Chart。命令如下。

```
helm create devops6-npm-service
replicaCount: 1
image:
  repository: 192.168.1.200:8088/devops6/devops6-npm-service
  pullPolicy: IfNotPresent
  tag: RELEASE-2.1.1-fbc1e7bb
```

将 Chart 代码提交到代码仓库。仓库内容如图 10-10 所示。

第 10 章 Kubernetes 持续部署

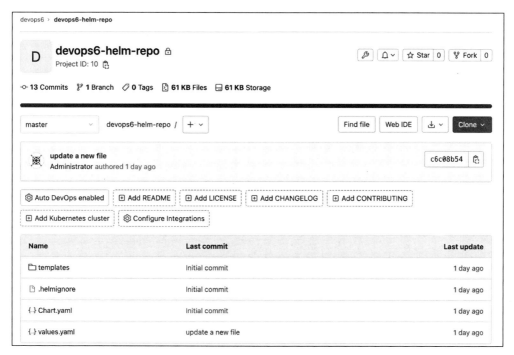

图 10-10 Kubernetes 部署清单仓库

Helm Chart 中的 values.yaml 文件存放的是一些变量配置信息。部分配置如下。

```
…
replicaCount: 1
image:
  repository: 192.168.1.200:8088/devops6/devops6-npm-service
  tag: RELEASE-2.1.1-fbc1e7bb
…
```

在持续集成流水线的更新镜像阶段，下载配置文件、替换镜像标签、更新文件。代码如下。

```
stage("UpdateEnvFile"){
    steps{
        script {
            projectId = 10              //Helm Chart 仓库项目地址
            fileName = "values.yaml"    //配置文件
            branchName = "master"

            //下载配置文件
            fileData = mygit.GetRepoFile(projectId,fileName, branchName )
            sh "rm -fr ${fileName}"
```

```
        //使用 readYaml 修改镜像 tag
        yamlData = readYaml text: fileData
        yamlData.image.tag = "${env.imageTag}"

        //将模板文件内容保存到本地
        writeYaml file: "${fileName}" , data: yamlData

        //创建/更新发布文件
        newYaml = sh returnStdout: true, script: "cat ${fileName}"
        println(newYaml)
        //更新 GitLab 文件内容
        base64Content = newYaml.bytes.encodeBase64().toString()

        //如果文件不存在则创建,存在则更新
        try {
            mygit.UpdateRepoFile(projectId,"${fileName}",base64Content,
"master")
        } catch(e){
            mygit.CreateRepoFile(projectId,"${fileName}",base64Content,
"master")
        }
    }
}
```

现在 Helm Chart 源代码存放在 Git 仓库中,也可以添加构建 Chart 阶段,将 Helm Chart 打包并上传到 Chart 仓库。流水线代码如下。

```
stage("HelmPackage"){
    steps{
        script{
            appName = "${JOB_NAME}".split('_')[0]
            sh "mkdir -p ${appName}"

            //下载 Chart 源码
            ws("${workspace}/${appName}"){
                checkout([$class: 'GitSCM', branches: [[name: '*/master']],
                    extensions: [],
                    userRemoteConfigs: [[credentialsId: 'gitlab-admin',
                    url: 'http://192.168.1.200:8076/devops6/devops6-helm-repo.git']]])
            }
            //将 Helm Chart 打包并上传到 Chart 仓库
```

```
            chartVersion = "${env.branchName}".split("-")[-1]    //获取版本号
            sh """
                helm package ${appName} --version ${chartVersion}
                helm cm-push ${appName}-${chartVersion}.tgz devops6repo
            """
        }
    }
}
```

10.2 基于 Kubectl 持续部署

基于 Kubernetes 环境实践持续部署流水线，该流水线包含下载清单文件、应用发布、人工验证、应用回滚阶段，如图 10-11 所示。

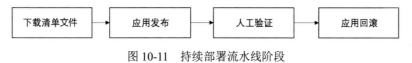

图 10-11 持续部署流水线阶段

10.2.1 准备工作

1. 创建 Pipeline

创建一个持续部署流水线作业，配置分支名称参数，如图 10-12 所示。

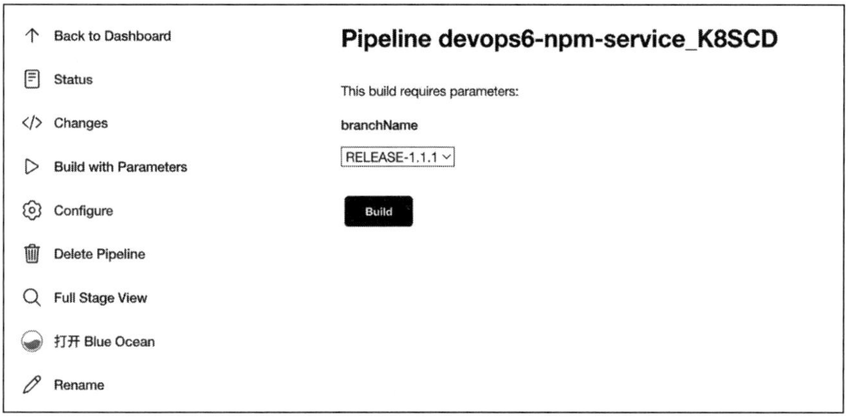

图 10-12 持续部署流水线参数

2. 配置 Jenkinsfile

选择加载共享库中的 Jenkinsfile。笔者将 Jenkinsfile 和共享库代码放到一起，持续部署流水线用的 Jenkinsfile 名称是 k8scd.jenkinsfile。SCM 选项用于选择仓库类型，这里选择 Git 类型，填写文件保存的远程 Git 仓库地址、认证凭据、分支名称、脚本路径等，如图 10-13 所示。

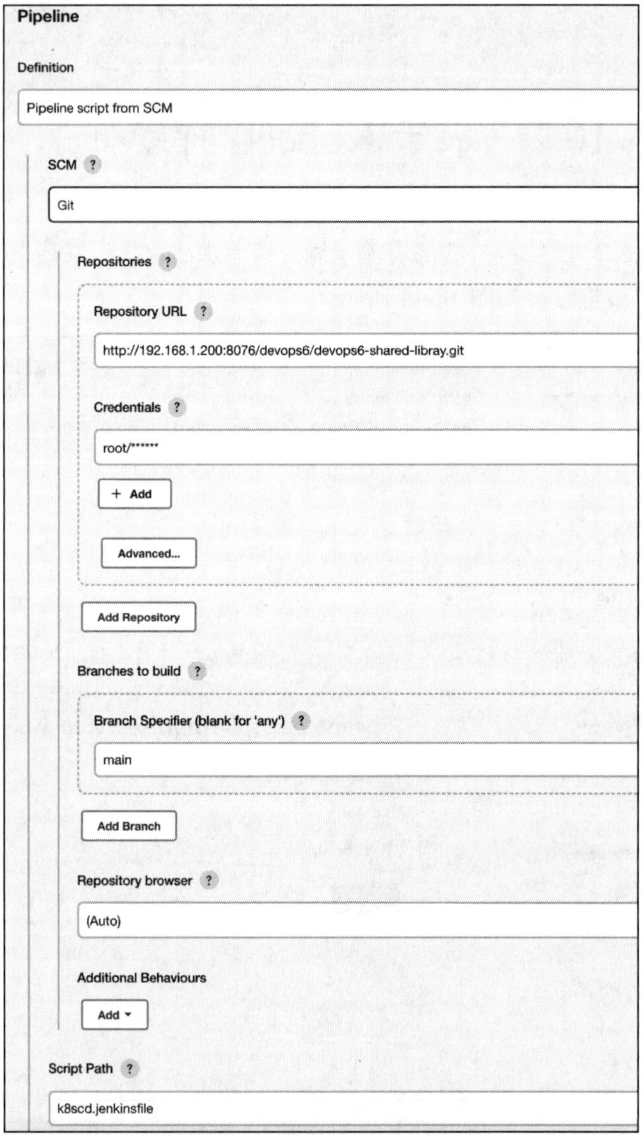

图 10-13　Jenkinsfile 脚本配置

3. 加载共享库

笔者在共享库中编写了一些关于持续集成的模块，Jenkinsfile 需要加载共享库并导入共享库中的模块。代码如下。

```
//加载共享库
@Library("devops06@main") _

//导入模块
def mygit = new org.devops.Gitlab()
```

10.2.2 设置 Pipeline

设置 Pipeline 运行的构建节点、运行时选项、持续部署的多个阶段。持续部署阶段分为 GetManifests（下载清单文件）、Deploy（应用发布）、RollBack（应用回滚）。代码如下。

```
//持续部署流水线
pipeline {
   agent { label "build" }
   options {
      skipDefaultCheckout true
   }
   stages{
      stage("GetManifests"){           //下载清单文件
         steps{
            script{
               println("GetManifests")
            }
         }
      }

      stage("Deploy"){                 //应用发布
         steps{
            script{
               println("Deploy")
            }
         }
      }

      stage("RollBack"){               //应用回滚
         input {
```

```
            message "是否进行回滚"
            ok "提交"
            submitter ""
            parameters {
                choice(choices: ['yes', 'no'], name: 'opts')
            }
        }
        steps{
            script{
                println("RollBack")
            }
        }
    }
}
```

1. 下载清单文件

通过 Jenkins 作业名称过滤出应用名称和版本分支，然后通过 GetRepoFile()函数下载清单文件仓库中 main 分支的部署文件并保存在本地。代码如下。

```
stage("GetManifests"){
    steps{
        script{
            projectId = 9                //清单文件仓库项目 ID
            env.deployFile = "${env.branchName}.yaml"
                                        //版本分支 RELEASE-1.1.1.yaml
            env.appName = "${JOB_NAME}".split('_')[0] //devops6-npm-service
            filePath = "${env.appName}%2f${env.deployFile}" //devops6-npm-service/RELEASE-2.1.1.yaml
            branchName = "main"         //分支名称
            fileData = mygit.GetRepoFile(projectId,filePath, branchName )
            sh "rm -fr ${env.deployFile}"
            writeFile file: env.deployFile , text: fileData
            sh "ls -l ; cat ${env.deployFile}"
        }
    }
}
```

下载清单文件阶段日志如图 10-14 所示。

```
[Pipeline] withCredentials
Masking supported pattern matches of $gitlabtoken
[Pipeline] {
[Pipeline] sh
Warning: A secret was passed to "sh" using Groovy String interpolation, which is insecure.
            Affected argument(s) used the following variable(s): [gitlabtoken]
            See https://jenkins.io/redirect/groovy-string-interpolation for details.
+ curl --location --request GET 'http://192.168.1.200:8076/api/v4//projects/9/repository/files/devops6-npm-service%2f
'PRIVATE-TOKEN: ****'
  % Total    % Received % Xferd  Average Speed   Time    Time     Time  Current
                                 Dload  Upload   Total   Spent    Left  Speed

  0     0    0     0    0     0      0      0 --:--:-- --:--:-- --:--:--     0
100  1030  100  1030    0     0  13205      0 --:--:-- --:--:-- --:--:-- 13205
[Pipeline] }
[Pipeline] // withCredentials
[Pipeline] readJSON
[Pipeline] echo
net.sf.json.JSONException: Invalid JSON String
[Pipeline] sh
+ rm -fr RELEASE-1.1.1.yaml
[Pipeline] writeFile
[Pipeline] sh
+ ls -l
总用量 4
-rw-r--r-- 1 root root 1030 5月  27 11:02 RELEASE-1.1.1.yaml
```

图 10-14　下载清单文件阶段日志

2．应用发布

使用 Kubectl 命令将应用发布到 Kubernetes 环境中。代码如下。

```
stage("Deploy"){
    steps{
        script{
            //获取名称空间
            env.namespace = "${env.appName}".split('-')[0]   //devops6

            ##发布应用
            sh """
                kubectl apply -f ${env.deployFile} -n ${env.namespace}
            """
            //获取应用状态
            5.times{
                sh "sleep 2; kubectl -n ${env.namespace} get pod | grep ${env.appName}"
            }
        }
    }
}
```

应用发布阶段日志如图 10-15 所示。

```
[Pipeline] sh
+ kubectl apply -f RELEASE-1.1.1.yaml -n devops6
deployment.apps/devops6-npm-service configured
service/devops6-npm-service unchanged
ingress.networking.k8s.io/devops6-npm-service unchanged
[Pipeline] sh
+ sleep 2
+ kubectl -n devops6 get pod
+ grep devops6-npm-service
devops6-npm-service-58b8d8bffb-f7j4x    1/1   Running             0   43m
devops6-npm-service-58b8d8bffb-fcprf    1/1   Running             0   43m
devops6-npm-service-6464557df9-72fkb    1/1   Running             0   3s
devops6-npm-service-6464557df9-9mt8f    0/1   ContainerCreating   0   2s
[Pipeline] sh
+ sleep 2
+ kubectl -n devops6 get pod
+ grep devops6-npm-service
devops6-npm-service-58b8d8bffb-f7j4x    1/1   Terminating         0   43m
devops6-npm-service-6464557df9-72fkb    1/1   Running             0   5s
devops6-npm-service-6464557df9-9mt8f    1/1   Running             0   4s
devops6-npm-service-6464557df9-nqb5n    1/1   Running             0   2s
[Pipeline] sh
+ sleep 2
+ kubectl -n devops6 get pod
+ grep devops6-npm-service
devops6-npm-service-6464557df9-72fkb    1/1   Running             0   7s
devops6-npm-service-6464557df9-9mt8f    1/1   Running             0   6s
devops6-npm-service-6464557df9-nqb5n    1/1   Running             0   4s
```

图 10-15　应用发布阶段日志

3. 应用回滚

可以使用 Kubectl 命令在 Kubernetes 环境中将应用回滚。代码如下。

```
stage("RollBack"){
    input {
        message "是否进行回滚"
        ok "提交"
        submitter ""
        parameters {
            choice(choices: ['yes', 'no'], name: 'opts')
        }
    }
    steps{
        script{
            switch("${opts}") {
                case "yes":
```

```
                    sh "kubectl rollout undo deployment/${env.appName} -n
${env.namespace} "
                    break
                case "no":
                    break
            }
        }
    }
}
```

应用回滚阶段日志如图 10-16 所示。

```
[Pipeline] // stage
[Pipeline] stage
[Pipeline] { (RollBack)
[Pipeline] input
Input requested
Approved by admin
[Pipeline] withEnv
[Pipeline] { (hide)
[Pipeline] script
[Pipeline] {
[Pipeline] sh
+ kubectl rollout undo deployment/devops6-npm-service -n devops6
deployment.apps/devops6-npm-service rolled back
[Pipeline] }
[Pipeline] // script
[Pipeline] }
[Pipeline] // withEnv
[Pipeline] }
[Pipeline] // stage
[Pipeline] }
[Pipeline] // node
[Pipeline] End of Pipeline
Finished: SUCCESS
```

图 10-16 应用回滚阶段日志

10.3 基于 Helm 持续部署

基于 Kubernetes 环境实践持续部署流水线，该流水线包含下载 Helm Chart、应用发布、人工验证、应用回滚阶段，如图 10-17 所示。

图 10-17　持续部署流水线阶段

10.3.1　准备工作

1．创建 Pipeline

创建一个持续部署流水线作业，配置分支名称参数，持续部署流水线参数如图 10-18 所示。

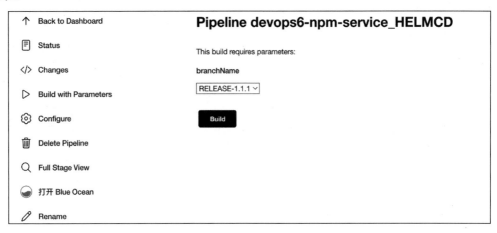

图 10-18　持续部署流水线参数

2．配置 Jenkinsfile

选择加载共享库中的 Jenkinsfile。笔者将 Jenkinsfile 和共享库代码放到一起，持续部署流水线用的 Jenkinsfile 名称是 helmcd.jenkinsfile。SCM 选项用于选择仓库类型，这里选择 Git 类型，填写文件保存的远程 Git 仓库地址、认证凭据、分支名称、脚本路径等，如图 10-19 所示。

3．加载共享库

笔者在共享库中编写了一些关于持续集成的模块，Jenkinsfile 需要加载共享库并导入共享库中的模块。代码如下。

```
//加载共享库
@Library("devops06@main") _

//导入模块
```

```
def checkout = new org.devops.Checkout()
def build = new org.devops.Build()
```

图 10-19　Jenkinsfile 脚本配置

10.3.2　设置 Pipeline

设置 Pipeline 运行的构建节点、运行时选项、持续部署的多个阶段。持续部署阶段分为

GetHelmChart（下载 Helm Chart）、Deploy（应用发布）、RollOut（应用回滚）。代码如下。

```
//持续部署流水线
pipeline {
   agent { label "build"}
   options {
      skipDefaultCheckout true
   }
   stages{
      stage("GetHelmChart"){          //下载 Helm Chart
         steps{
            script{
               println("GetHelmChart")
            }
         }
      }

      stage("Deploy"){                //应用发布
         steps{
            script{
               println("Deploy")
            }
         }
      }
      stage("RollOut"){               //应用回滚
         input {
            message "是否进行回滚"
            ok "提交"
            submitter ""
            parameters {
               choice(choices: ['yes', 'no'], name: 'opts')
            }
         }
         steps{
            script{
               println("RollOut")
            }
         }
      }
   }
}
```

1. 下载 Helm Chart

通过 Jenkins 作业名称过滤出应用名称和版本分支，然后通过 GetRepoFile() 函数下载清单文件仓库中 main 分支中的部署文件并保存在本地。代码如下。

```
stage("GetHelmChart"){
    steps{
        script{
            //下载helm chart
            env.chartVersion = "${env.branchName}".split("-")[-1]
            env.appName = "${JOB_NAME}".split('_')[0]
            sh """
                helm repo update devops6repo
                helm pull devops6repo/${env.appName} --version ${env.chartVersion}
            """
        }
    }
}
```

下载 Helm Chart 阶段日志如图 10-20 所示。

```
RELEASE-2.1.1
[Pipeline] node
Running on build01 in /opt/jenkinsagent/workspace/devops6-npm-service_HELMCD@2
[Pipeline] {
[Pipeline] stage
[Pipeline] { (GetHelmChart)
[Pipeline] script
[Pipeline] {
[Pipeline] sh
+ helm repo update devops6repo
Hang tight while we grab the latest from your chart repositories...
...Successfully got an update from the "devops6repo" chart repository
Update Complete. *Happy Helming!*
+ helm pull devops6repo/devops6-npm-service --version 2.1.1
[Pipeline] }
[Pipeline] // script
[Pipeline] }
[Pipeline] // stage
[Pipeline] stage
[Pipeline] { (Deploy)
[Pipeline] script
```

图 10-20　下载 Helm Chart 阶段日志

2. 应用发布

使用 Helm 命令将应用发布到 Kubernetes 环境中。代码如下。

```groovy
stage("Deploy"){
    steps{
        script{
            env.namespace = "${env.appName}".split('-')[0]    //devops6
            //发布应用
            sh """
                helm upgrade --install --create-namespace "${env.appName}" ./"${env.appName}"-${env.chartVersion}.tgz -n ${env.namespace}
                helm history "${env.appName}" -n ${env.namespace}
            """

            //获取应用状态
            5.times{
                sh "sleep 2; kubectl -n ${env.namespace} get pod | grep ${env.appName}"
            }

            //收集历史版本,便于回滚
            env.revision = sh returnStdout: true,
                    script: """helm history ${env.appName} -n ${env.namespace} | grep -v 'REVISION' | awk '{print \$1}' """
            println("${env.revision}")
            println("${env.revision.split('\n').toString()}")
            env.REVISION = "${env.revision.split('\n').toString()}"
            println("${env.REVISION}")
        }
    }
}
```

应用发布阶段日志如图 10-21 所示。

```
[Pipeline] sh
+ helm upgrade --install --create-namespace devops6-npm-service ./devops6-npm-service-2.1.1.tgz -n devops6
Release "devops6-npm-service" has been upgraded. Happy Helming!
NAME: devops6-npm-service
LAST DEPLOYED: Sat May 27 12:18:49 2023
NAMESPACE: devops6
STATUS: deployed
REVISION: 3
NOTES:
1. Get the application URL by running these commands:
   http://devops.test.com/
+ helm history devops6-npm-service -n devops6
REVISION    UPDATED                     STATUS      CHART                       APP VERSION    DESCRIPTION
1           Sat May 27 12:15:20 2023    superseded  devops6-npm-service-2.1.1   1.16.0         Install complete
2           Sat May 27 12:17:09 2023    superseded  devops6-npm-service-1.1.1   1.16.0         Upgrade complete
3           Sat May 27 12:18:49 2023    deployed    devops6-npm-service-2.1.1   1.16.0         Upgrade complete
[Pipeline] sh
+ sleep 2
```

图 10-21 应用发布阶段日志

3. 应用回滚

使用 Helm 命令在 Kubernetes 环境中将应用回滚。这里使用应用发布阶段获取的 Helm 应用版本，回滚的时候可以交互式指定要回滚的版本号。代码如下。

```
stage("RollOut"){
  input {
     message "是否进行回滚"
     ok "提交"
     submitter ""
     parameters {
        choice(choices: ['yes', 'no'], name: 'opts')
     }
  }


  steps{
     script{

        switch("${opts}") {
           case "yes":
           def result = input  message: "选择回滚版本",
              parameters:[choice(choices:env.REVISION,name:'rversion')]

           println("${result}")
           sh "helm rollback ${env.appName} ${result} -n ${env.namespace} "
           break

           case "no":
           break
        }
     }
  }
}
```

应用回滚阶段日志如图 10-22 所示。

```
[Pipeline] { (RollOut)
[Pipeline] input
Input requested
Approved by admin
[Pipeline] withEnv
[Pipeline] {
[Pipeline] script
[Pipeline] {
[Pipeline] input
Input requested
Approved by admin
[Pipeline] echo
2
[Pipeline] sh
+ helm rollback devops6-npm-service 2 -n devops6
Rollback was a success! Happy Helming!
[Pipeline] }
[Pipeline] // script
[Pipeline] }
[Pipeline] // withEnv
[Pipeline] }
[Pipeline] // stage
[Pipeline] }
[Pipeline] // node
[Pipeline] End of Pipeline
Finished: SUCCESS
```

图 10-22　应用回滚阶段日志

10.4　本章小结

本章我们实践了基于 Kubernetes 的持续部署实践方式。读者可以掌握基于 GitOps 的实践方式设计持续集成和持续部署流水线，即分别通过 Kubectl 和 Helm 工具实现，进而了解原生的应用发布过程。

第 11 章
基础设施即代码

Terraform 是一款开源的自动化部署和配置工具，用于创建、部署和管理云计算基础设施。它可以轻松地构建、部署和管理云基础架构，同时也支持使用 Provider 在任何环境中进行自动化部署和配置。

本章内容速览。
- ☑ Terraform 入门
- ☑ 供应商 Provider
- ☑ 定义云资源
- ☑ 开通资源
- ☑ 本章小结

11.1 Terraform 入门

1. Terraform 概述

Terraform 是一种基础设施即代码工具，可以在配置文件中定义云和本地资源并对其进行版本控制、重用和共享，还可以使用一致的工作流程在整个生命周期内配置和管理所有基础架构。Terraform 可以管理的资源有 DNS 记录、SaaS 功能、计算、存储和网络资源等。以下代码定义了一个阿里云实例。

```
resource "alicloud_instance" "myecs" {
  availability_zone        = var.region
  security_groups          = [var.secgroup_id]
  instance_type            = var.instance_type
  system_disk_category     = "cloud_efficiency"
  system_disk_name         = "tf_system_disk_name"
  system_disk_description  = "tf_system_disk_description"
```

```
    image_id                         = data.alicloud_images.images_ds.images[0].id
    instance_name                    = var.instance_name
    vswitch_id                       = var.vsw_id
    internet_max_bandwidth_out       = 0
    internet_charge_type             = "PayByTraffic"
    password                         = "root@123"
    user_data = <<-EOF
            #!/bin/bash
            # until [[ -f /var/lib/cloud/instance/boot-finished ]] ;do
            #    sleep 1
            # done
            yum -y install nginx
            echo "myserver" >/usr/share/nginx/html/index.html
            systemctl restart nginx
            EOF
}
```

Terraform 通过其应用程序编程接口（API）在云平台和其他服务上创建和管理资源。Provider 使 Terraform 能够通过可访问的 API 与几乎任何平台或服务一起工作，Terraform 工作原理如图 11-1 所示。

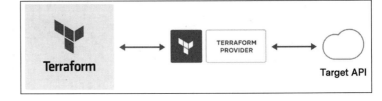

图 11-1　Terraform 工作原理

Terraform 的配置文件以 .tf 为后缀。Terraform 支持两种语法模式 HCL、JSON，HCL 即 HashiCorp Configuration Language，这里笔者使用常用的 HCL 语法演示。

2．安装 Terraform

Terraform 可以支持安装在各个平台。访问 https://www.terraform.io/downloads 进入 Terraform 的安装包下载页面，如图 11-2 所示。

笔者使用的是 Linux 平台，安装命令如下。

```
#解压安装包
unzip terraform_1.1.7_darwin_amd64.zip
#将可执行程序移动到系统变量
mv terraform /usr/local/bin
#验证安装成功
terraform -version
```

第 11 章 基础设施即代码

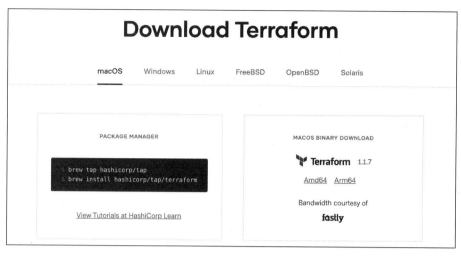

图 11-2　Terraform 的安装包下载页面

11.2　供应商 Provider

Terraform 通过 Provider 管理基础设施，使用 Provider 与云供应商 API 进行交互。每个 Provider 都包含相关的资源和数据源。官方链接为 https://registry.terraform.io/providers，Terraform 支持的 Providers 如图 11-3 所示。

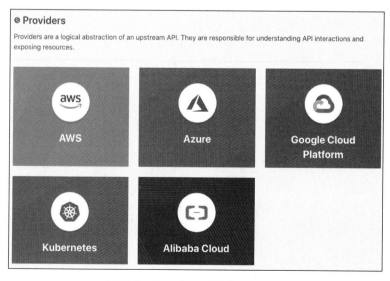

图 11-3　Terraform 支持的 Providers

1. 声明 Provider

每个 Terraform 模块必须声明它需要哪些 Provider，以便 Terraform 可以安装和使用它们。创建 versions.tf 文件，然后配置使用阿里云的 Provider，文件内容如下。

```
terraform {
  required_version = "1.1.7"
  required_providers {
    alicloud = {
      //源地址
      source = "aliyun/alicloud"
      //版本
      version = "1.162.0"
    }
  }
}
```

关于每个 Provider 可以参考官方对应的文档，例如阿里云的 Provider 使用文档（https://registry.terraform.io/providers/aliyun/alicloud/latest），如图 11-4 所示。

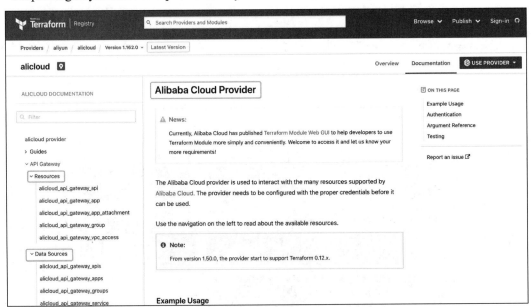

图 11-4　阿里云的 Provider 使用文档

2. 配置 Provider

通常每个云供应商的 Provider 都需要配置操作资源的地域（region）信息，以及用户认证 access_key 和 secret_key 信息。这些信息可以在云供应商的控制台获取。

进入阿里云控制台，创建用户，如图 11-5 所示。

图 11-5　创建用户

获取当前用户的 access_key 和 secret_key 信息，如图 11-6 所示。

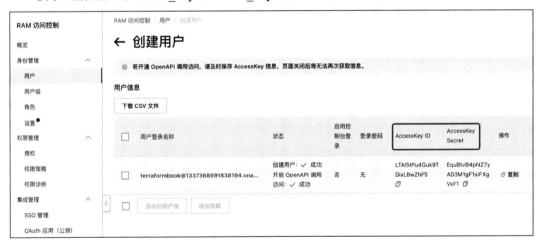

图 11-6　获取用户的 access_key 和 secret_key 信息

Provider 的配置代码如下。

```
provider "alicloud" {
  region = "cn-beijing"
  access_key = <access_key>
  secret_key = <secret_key>
}
```

11.3 定义云资源

Resource（资源）来自 Provider，是 Terraform 中最重要的元素。每个资源块描述一个或多个基础对象，如网络、计算实例或更高级别的组件（如 DNS 记录）。

1. 资源语法

根据语法，资源名称必须以字母或下画线开头，并且只能包含字母、数字、下画线和破折号。语法格式如下。

```
resource "resource_type" "name" {
  //resource_config
}
```

定义一个阿里云上的 DNS 记录资源，域名是 demo.devops.site，记录类型为 A，值为 IP 地址 123.123.123.123。代码如下。

```
resource "alicloud_dns_record" "record" {
  name = "devops.site"              //zone 名称
  host_record = "demo"              //主机记录
  type = "A"                        //记录类型
  value = "123.123.123.123"         //记录值
}
```

2. 专有网络

专有网络 VPC（virtual private cloud）是用户基于阿里云创建的自定义私有网络。定义一个 VPC 专有网络 tf_test 和一个 vswitch 虚拟交换机。代码如下。

```
//VPC 专有网络
resource "alicloud_vpc" "vpc" {
  vpc_name   = "tf_test"
  cidr_block = "172.80.0.0/12"
}

//switch 交换机
resource "alicloud_vswitch" "vsw" {
  vpc_id     = alicloud_vpc.vpc.id
  cidr_block = "172.80.0.0/21"
  zone_id    = "cn-beijing-b"
}
```

3. 安全组

安全组是一种虚拟防火墙，用于控制安全组内实例的入流量和出流量，在控制台创建安全组页面如图 11-7 所示。

图 11-7 控制台创建安全组页面

定义安全组的代码如下。

```
//security_group 安全组
resource "alicloud_security_group" "group" {
  name               = "demo-group"
  vpc_id             = alicloud_vpc.vpc.id
  security_group_type = "normal"    //普通类型
}
```

定义安全组中出/入流量规则的代码如下。

```
//security_group_rule 规则
resource "alicloud_security_group_rule" "allow_80_tcp" {
  type        = "ingress"
  ip_protocol = "tcp"
  nic_type    = "intranet"
  policy      = "accept"
  port_range  = "80/80"
```

```
  priority             = 1
  security_group_id = alicloud_security_group.group.id
  cidr_ip             = "0.0.0.0/0"
}

//security_group_rule 规则
resource "alicloud_security_group_rule" "allow_22_tcp" {
  type                = "ingress"
  ip_protocol         = "tcp"
  nic_type            = "intranet"
  policy              = "accept"
  port_range          = "22/22"
  priority            = 1
  security_group_id = alicloud_security_group.group.id
  cidr_ip             = "0.0.0.0/0"
}
```

11.4 开通资源

Terraform 的工作流程主要分为 3 个阶段：Write、Plan、Apply。Write 阶段，定义所需要的资源，这些资源可以是跨越多个云提供商和服务。例如，定义一个具有安全组和负载均衡器的 ECS 机器。Plan 阶段，Terraform 将创建一个执行计划，根据配置文件以命令行输出的方式计算出要创建、更新或销毁的基础设施资源。例如，可以查看当前基础设施并与期望状态进行对比，便于发布前对配置进行预览。Apply 阶段，在批准后，Terraform 会按照资源依赖关系顺序执行资源创建。

1. 初始化

Terraform 初始化是使用 Terraform 工具的第一步，它用于设置和准备 Terraform 环境。在初始化过程中，Terraform 安装 Provider、模块，配置后端 State。.terraform 目录在初始化时自动创建，Terraform 使用它来管理缓存的提供程序插件和模块等配置。state 文件用于存储 Terraform 的数据——terraform.tfstate。初始化过程的日志如图 11-8 所示。

2. 格式化与验证

为保证 Terraform 代码整洁，可以通过 terraform fmt 命令格式化代码，如图 11-9 所示。terraform validate 命令可以对定义的资源代码进行语法校验，如图 11-10 所示。

```
→ terraform-example terraform init

Initializing the backend...

Initializing provider plugins...
- Finding aliyun/alicloud versions matching "1.162.0"...
- Installing aliyun/alicloud v1.162.0...
- Installed aliyun/alicloud v1.162.0 (signed by a HashiCorp partner, key ID 47422B
4AA9FA381B)

Partner and community providers are signed by their developers.
If you'd like to know more about provider signing, you can read about it here:
https://www.terraform.io/docs/cli/plugins/signing.html

Terraform has created a lock file .terraform.lock.hcl to record the provider
selections it made above. Include this file in your version control repository
so that Terraform can guarantee to make the same selections by default when
you run "terraform init" in the future.

Terraform has been successfully initialized!
```

图 11-8　Terraform 初始化过程的日志

```
→ terraform-example terraform fmt
alicloud_security_group.tf
main.tf
versions.tf
```

图 11-9　Terraform 格式化代码

```
→ terraform-example terraform validate -json
{
  "format_version": "1.0",
  "valid": true,
  "error_count": 0,
  "warning_count": 0,
  "diagnostics": []
}
```

图 11-10　Terraform 语法验证

3. 预览资源

评估 Terraform 配置，并打印声明的所有资源的期望状态，将期望状态与当前工作目录的基础设施对象进行比较。打印当前状态和期望状态之间的差异（不会执行变更），Terraform 资源预览如图 11-11 所示。

4. 发布资源

terraform apply 命令可以运行 plan（计划）中的操作进行资源开通，如图 11-12 所示。

销毁资源可以使用 terraform destroy 命令。在生产环境中应谨慎使用此命令，避免资源删除导致故障，如图 11-13 所示。

```
→ terraform-example terraform plan
Terraform used the selected providers to generate the following execution plan.
Resource actions are indicated with the following symbols:
  + create

Terraform will perform the following actions:

  # alicloud_security_group.group will be created
  + resource "alicloud_security_group" "group" {
      + id                  = (known after apply)
      + inner_access        = (known after apply)
      + inner_access_policy = (known after apply)
      + name                = "demo-group"
      + security_group_type = "normal"
      + vpc_id              = (known after apply)
    }

  # alicloud_security_group_rule.allow_80_tcp will be created
  + resource "alicloud_security_group_rule" "allow_80_tcp" {
      + cidr_ip           = "0.0.0.0/0"
      + id                = (known after apply)
      + ip_protocol       = "tcp"
      + nic_type          = "internet"
      + policy            = "accept"
      + port_range        = "80/80"
      + prefix_list_id    = (known after apply)
      + priority          = 1
      + security_group_id = (known after apply)
      + type              = "ingress"
    }

  # alicloud_vpc.vpc will be created
  + resource "alicloud_vpc" "vpc" {
      + cidr_block        = "172.16.0.0/12"
      + id                = (known after apply)
      + ipv6_cidr_block   = (known after apply)
      + name              = (known after apply)
      + resource_group_id = (known after apply)
      + route_table_id    = (known after apply)
      + router_id         = (known after apply)
      + router_table_id   = (known after apply)
      + status            = (known after apply)
      + vpc_name          = "tf_test"
    }

  # alicloud_vswitch.vsw will be created
  + resource "alicloud_vswitch" "vsw" {
      + availability_zone = (known after apply)
      + cidr_block        = "172.16.0.0/21"
      + id                = (known after apply)
      + name              = (known after apply)
      + status            = (known after apply)
      + vpc_id            = (known after apply)
      + vswitch_name      = (known after apply)
      + zone_id           = "cn-beijing-b"
    }

Plan: 4 to add, 0 to change, 0 to destroy.
```

图 11-11　Terraform 资源预览

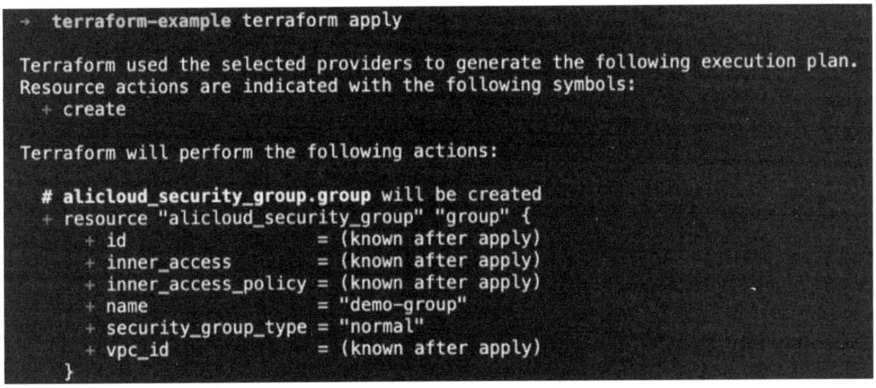

图 11-12 Terraform 开通资源

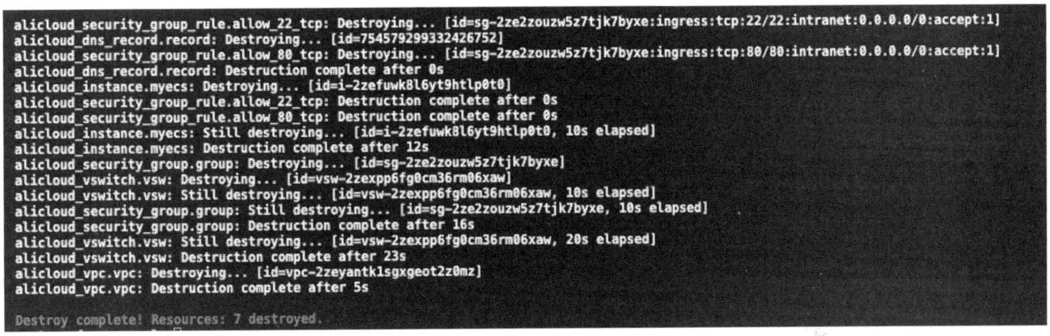

图 11-13 Terraform 销毁资源

11.5 本章小结

Terraform 是一款功能强大且易于使用的工具，使用不同的 Provider 可以实现多云环境下的云架构资源管理并且可以灵活地适应各种不同类型的业务需求。